Springer Tracts in Modern Physics
Volume 187

Managing Editor: G. Höhler, Karlsruhe

Editors: J. Kühn, Karlsruhe
Th. Müller, Karlsruhe
A. Ruckenstein, New Jersey
F. Steiner, Ulm
J. Trümper, Garching
P. Wölfle, Karlsruhe

Now also Available Online

Starting with Volume 165, Springer Tracts in Modern Physics is part of the [SpringerLink] service. For all customers with standing orders for Springer Tracts in Modern Physics we offer the full text in electronic form via [SpringerLink] free of charge. Please contact your librarian who can receive a password for free access to the full articles by registration at:

www.springerlink.com/orders/index.htm

If you do not have a standing order you can nevertheless browse through the table of contents of the volumes and the abstracts of each article at:

www.springerlink.com/series/stmp/

There you will also find more information about the series.

Springer
Berlin
Heidelberg
New York
Hong Kong
London
Milan
Paris
Tokyo

Physics and Astronomy — ONLINE LIBRARY

http://www.springer.de/phys/

Springer Tracts in Modern Physics

Springer Tracts in Modern Physics provides comprehensive and critical reviews of topics of current interest in physics. The following fields are emphasized: elementary particle physics, solid-state physics, complex systems, and fundamental astrophysics.

Suitable reviews of other fields can also be accepted. The editors encourage prospective authors to correspond with them in advance of submitting an article. For reviews of topics belonging to the above mentioned fields, they should address the responsible editor, otherwise the managing editor. See also http://www.springer.de/phys/books/stmp.html

Managing Editor

Gerhard Höhler

Institut für Theoretische Teilchenphysik
Universität Karlsruhe
Postfach 69 80
76128 Karlsruhe, Germany
Phone: +49 (7 21) 6 08 33 75
Fax: +49 (7 21) 37 07 26
Email: gerhard.hoehler@physik.uni-karlsruhe.de
http://www-ttp.physik.uni-karlsruhe.de/

Elementary Particle Physics, Editors

Johann H. Kühn

Institut für Theoretische Teilchenphysik
Universität Karlsruhe
Postfach 69 80
76128 Karlsruhe, Germany
Phone: +49 (7 21) 6 08 33 72
Fax: +49 (7 21) 37 07 26
Email: johann.kuehn@physik.uni-karlsruhe.de
http://www-ttp.physik.uni-karlsruhe.de/~jk

Thomas Müller

Institut für Experimentelle Kernphysik
Fakultät für Physik
Universität Karlsruhe
Postfach 69 80
76128 Karlsruhe, Germany
Phone: +49 (7 21) 6 08 35 24
Fax: +49 (7 21) 6 07 26 21
Email: thomas.muller@physik.uni-karlsruhe.de
http://www-ekp.physik.uni-karlsruhe.de

Fundamental Astrophysics, Editor

Joachim Trümper

Max-Planck-Institut für Extraterrestrische Physik
Postfach 16 03
85740 Garching, Germany
Phone: +49 (89) 32 99 35 59
Fax: +49 (89) 32 99 35 69
Email: jtrumper@mpe-garching.mpg.de
http://www.mpe-garching.mpg.de/index.html

Solid-State Physics, Editors

Andrei Ruckenstein
Editor for The Americas

Department of Physics and Astronomy
Rutgers, The State University of New Jersey
136 Frelinghuysen Road
Piscataway, NJ 08854-8019, USA
Phone: +1 (732) 445 43 29
Fax: +1 (732) 445-43 43
Email: andreir@physics.rutgers.edu
http://www.physics.rutgers.edu/people/pips/Ruckenstein.html

Peter Wölfle

Institut für Theorie der Kondensierten Materie
Universität Karlsruhe
Postfach 69 80
76128 Karlsruhe, Germany
Phone: +49 (7 21) 6 08 35 90
Fax: +49 (7 21) 69 81 50
Email: woelfle@tkm.physik.uni-karlsruhe.de
http://www-tkm.physik.uni-karlsruhe.de

Complex Systems, Editor

Frank Steiner

Abteilung Theoretische Physik
Universität Ulm
Albert-Einstein-Allee 11
89069 Ulm, Germany
Phone: +49 (7 31) 5 02 29 10
Fax: +49 (7 31) 5 02 29 24
Email: steiner@physik.uni-ulm.de
http://www.physik.uni-ulm.de/theo/theophys.html

Karl Leo

High-Field Transport in Semiconductor Superlattices

With 164 Figures

 Springer

Professor Dr. Karl Leo
Institut für Angewandte Photophysik
Technische Universität Dresden
01062 Dresden, Germany
E-mail: leo@iapp.de
http://www.karl-leo.de

Library of Congress Cataloging-in-Publication Data
Leo, Karl.
High-field transport in semiconductor superlattices / Karl Leo.
p. cm. - - (Springer tracts in modern physics, ISSN 0081-3869 ; v. 187)
Includes bibliographical references and index.
ISBN 3-540-00569-2 (acid-free paper)
1. Electron transport. 2. Superlattices as materials. 3. Semiconductors. I. Title. II. Series. III. Springer tracts in modern physics ; 187.

QC1.S797 vol. 187
539 s- -dc21
[537.6'22]   20030503900

Physics and Astronomy Classification Scheme (PACS):
72.20.-i, 72.20.Ht, 73.21.Cd, 78.47.+p, 78.67.Pt

ISSN print edition: 0081-3869
ISSN electronic edition: 1615-0430
ISBN 3-540-00569-2 Springer-Verlag Berlin Heidelberg New York

This work is subject to copyright. All rights are reserved, whether the whole or part of the material is concerned, specifically the rights of translation, reprinting, reuse of illustrations, recitation, broadcasting, reproduction on microfilm or in any other way, and storage in data banks. Duplication of this publication or parts thereof is permitted only under the provisions of the German Copyright Law of September 9, 1965, in its current version, and permission for use must always be obtained from Springer-Verlag. Violations are liable for prosecution under the German Copyright Law.

Springer-Verlag Berlin Heidelberg New York
a member of BertelsmannSpringer Science+Business Media GmbH

http://www.springer.de

© Springer-Verlag Berlin Heidelberg 2003
Printed in Germany

The use of general descriptive names, registered names, trademarks, etc. in this publication does not imply, even in the absence of a specific statement, that such names are exempt from the relevant protective laws and regulations and therefore free for general use.

Typesetting: Camera-ready copy from the author using a Springer LATEX macro package
Production: LE-TEX Jelonek, Schmidt & Vöckler GbR, Leipzig
Cover concept: eStudio Calamar Steinen
Cover production: *design & production* GmbH, Heidelberg

Printed on acid-free paper   SPIN: 10770584   56/3141/YL   5 4 3 2 1 0

Für Elke, Ulrich und Karoline

Preface

It is obvious that carrier transport in solids is one of the most important topics in physics due to its ubiquitous applications in modern-day life, with microelectronics being a key example. After the pioneering work of Drude, the understanding of solid state transport made major advances when quantum mechanics was developed. Key contributions by Bloch and Zener laid the foundation for all further work in this area.

However, already in the early work of Bloch and Zener, peculiar effects in the high-field region were predicted: At high fields, electrons should oscillate instead of drifting along the field. Also, they should be transferred to higher bands, leading to, e.g., electronic breakdown. For many decades, these high-field effects were more a curiosity and matter of theoretical debates.

With the advent of the semiconductor superlattice in 1970, this situation changed dramatically: It turned out that these structures were an ideal playground within which to investigate high-field transport. Accordingly, Bloch oscillations were first observed in 1992, nearly 60 years after they were predicted. Since then, a large number of experimental and theoretical studies on this and related topics have appeared. Many groups worldwide work on gaining a better understanding of the basic physics and on possible novel devices using these high-field transport effects.

There have been a number of reviews of the field of semiconductor superlattices in general [1], their optical [2] and transport properties [3,4], and on Bloch oscillations [5–10]. The aim of this book is to give a concise introduction to this field and to summarize the important recent developments. Due to the rapid progress of the field, this can only be a limited overview of the field. Furthermore, I try to discuss the lines along which future research and development of device concepts could go.

In the course of doing research on Bloch oscillations, I have collaborated with many people. It is a pleasure for me to acknowledge many colleagues: In historical order, I would like to start with Jagdeep Shah, who first suggested that I ought investigate Bloch oscillations in the spring of 1990, and Ernst Göbel, who guided my first steps in nonlinear spectroscopy of coherent excitations. Later on, all significant results were obtained with samples grown by Klaus Köhler, whose contributions which went well beyond simple "sample delivery" were crucial. I enjoyed the collaboration with Jochen Feldmann at Bell Labs. In Aachen and Dresden, key experimental contributions were made by Peter Haring Bolivar (who obtained the first unambiguous signature of the oscillations), Patrick Leisching, who did many experiments which he previously had declared to be impossible, Vadim Lyssenko, master of the optics lab, who found the Wannier–Stark ladder peak shifts, Falk Löser, who sucessfully used those for elegant transport experiments, Markas Sudzius, incredibly patient ("Professor, this experiment does not work") but finally successful, and Ben Rosam and Dirk Meinhold for bringing Zener tunneling into the world of Bloch oscillations, using very carefully done experiments. Key theoretical support came from Marc Dignam, who carefully answered all the silly questions only experimentalists can ask.

For further experimental and theoretical collaboration, I would like to thank Friedhelm Bechstedt, Wolfgang Beck, Jack Cunningham, Thomas Dekorsy, Youssef Dhaibi, Peter Ganser, Stephan Glutsch, Tom Hasche, Christian Holfeld, Yuri Kosevich, Heinrich Kurz, Karl-Heinz Pantke, Gero von Plessen, Hartmut Roskos, Fausto Rossi, Wilfried Schäfer, the late Stefan Schmitt-Rink, Ralf Schwedler, Peter Thomas, Gintaras Valusis, Christian Waschke, Jianzhong Zhang, and many others who cannot all be mentioned here.

I would like to thank Manfred Helm and Harald Schneider for a careful reading of the manuscript and for many helpful comments, and Dirk Meinhold, Ben Rosam, and Rico Schüppel for their meticulous technical help with the manuscript. I would also like to thank Mark Fox and David Miller for their permission to use the transfer matrix software TRMF, which has been of enormous help in designing many of the sample structures which are discussed in this book.

I have also profited from many discussions with colleagues over this topic: To name a few, I would like to thank Daniel Chemla, David Dunlap, Leo Esaki, Erich Gornik, Herbert Kroemer, Claude Salomon, and Ray Tsu for their insight which helped me to, at least partially, understand what Bloch oscillations and Zener tunneling really are...

Finally, I thank my teachers Adolf Goetzberger, Joachim Knobloch, Hans Queisser, Wolfgang Rühle, Roland Schindler, and Bernhard Voss for introducing me to research in semiconductors.

Dresden, June 2003 *Karl Leo*

References

1. *Semiconductor Superlattices – Growth and Electronic Properties*, Ed. H.T. Grahn (World Scientific, Singapore 1995).
2. M. Helm, Semic. Sci. Technol. **10**, 557 (1995).
3. A. Wacker, Phys. Rep. **357**, 1 (2002).

4. A. Sibille, C. Minot, and F. Laruelle, Int. J. of Mod. Phys. B **14**, 909 (2000).
5. F. Rossi, Semic. Sci. Technol. **13**, 147 (1998).
6. K. Leo, Semic. Sci. Technol. **13**, 249 (1998).
7. F. Rossi, *Bloch Oscillations and Wannier–Stark Localization in Semiconductor Superlattices*, in *Theory of Transport Properties of Semiconductor Nanostructures*, Ed. E. E. Schöll, Chapman and Hall, London 1998.
8. V.G. Lyssenko, M. Sudzius, F. Löser, G. Valusis, T. Hasche, K. Leo, M.M. Dignam, and K. Köhler, Festkörperprobleme/Advances in Solid State Physics **38**, 225 (1999).
9. K. Leo and M. Koch, in *The Physics and Chemistry of Wave Packets*, Ed. J. Yeazell and T. Uzer (John Wiley, New York 2000), p. 285.
10. P. Haring Bolivar, T. Dekorsy, and H. Kurz, in *Semiconductors and Semimetals*, Vol. 66, ed. G. Mueller (Academic Press, San Diego 2000), p. 187.

Contents

1 Introduction: Basics of High Field Transport 1
   1.1 Introductory Remarks 1
   1.2 Coherent vs. Incoherent Transport of Electrons in a Band ... 2
   1.3 Basics of Bloch Oscillations 4
   1.4 Coupling to Higher Bands: Zener Tunneling 7
   References 8

2 Semiconductor Superlattices 9
   2.1 Introduction 9
   2.2 Electronic States in Superlattices Without an Electric Field .. 11
   2.3 Electronic States in Superlattices with an Electric Field 15
      2.3.1 Introduction 15
      2.3.2 The Kane Approach 16
      2.3.3 The Tight-Binding Approach 18
   2.4 Optical Transitions in Biased Superlattices 20
      2.4.1 General Picture 20
      2.4.2 The Excitonic Wannier–Stark Ladder 21
      2.4.3 Experimental Observation
           of the Wannier–Stark Ladder 23
   2.5 Summary 25
   References 25

**3 Interband Optical Experiments on Bloch Oscillations:
Basic Observations** 27
   3.1 Coherent Optical Excitation of Superlattices:
      Relation to the Gedankenexperiment 27
   3.2 Interband vs. Intraband Dynamics in Optical Experiments ... 29
      3.2.1 Introduction 29
      3.2.2 Description of Bloch Oscillations 32
      3.2.3 Extension to Real Semiconductors 33
   3.3 Optical Techniques to Detect Bloch Oscillations 35
      3.3.1 Interband Optical Techniques 35
      3.3.2 Other Techniques 41
   3.4 Observation of Bloch Oscillations
      by Transient Four-Wave Mixing 42

	3.5	Further Investigations with Four-Wave Mixing	47
		3.5.1 Quantum vs. Polarization Oscillations	47
		3.5.2 Influence of Miniband Width	48
	3.6	Summary	49
		References	50

4 Interband Optical Experiments on Bloch Oscillations: Measurements of the Spatial Dynamics 53
  4.1 Determination of the Spatial Amplitude from Transient Peak Shifts 53
  4.2 Control of the Amplitude by Changing the Laser Excitation: Tuning Between Breathing Modes and Spatial Oscillations ... 60
  4.3 Influence of Scattering and Coupling to the Plasma on Bloch Oscillation Dynamics 67
  4.4 Self-Induced Shapiro Effect 76
     4.4.1 Relation Between Bloch Oscillations and Josephson Effect 76
     4.4.2 Observation of Linear Transport in Optically Excited Superlattices 77
     4.4.3 Theoretical Model 80
  4.5 Summary 85
  References 85

5 Emission of Terahertz Radiation 87
  5.1 Techniques of Terahertz Spectroscopy 87
  5.2 Basic Experiments Observing THz Emission from Bloch Oscillations 89
  5.3 Temperature Dependence 91
  5.4 Excitation Above the Band Edge 92
  5.5 THz Experiments on Miniband Transport and Crossover to Bloch Oscillations 97
  5.6 THz Emission in Crossed Electric and Magnetic Fields 100
  5.7 Summary 101
  References 102

6 Damping of Bloch Oscillations I: Scattering 103
  6.1 Damping Times in Optical and Transport Experiments 103
  6.2 Damping of Bloch Oscillations by Phonon and Interface Scattering 105
     6.2.1 Decay Dynamics vs. Lattice Temperature 106
     6.2.2 Decay Dynamics vs. Excitation Intensity 109
     6.2.3 Discussion of the Decay Dynamics 111
  6.3 Coupling of Bloch Oscillations and Optical Phonons 113
  6.4 Summary 116
  References 116

7 Damping of Bloch Oscillations II: Zener Tunneling 119
7.1 Spectrum of a Superlattice in a Very High Field 121
- 7.1.1 First Experiments 121
- 7.1.2 Theoretical Considerations 123
- 7.1.3 Experimental Investigation of Zener Tunneling in Linear Optical Experiments 125
- 7.1.4 Dynamics of Single Wannier–Stark Ladder States in the Presence of Strong Zener Tunneling 134

7.2 Wave Packet Dynamics in the Zener Tunneling Regime 139
- 7.2.1 Experimental Approach 139
- 7.2.2 Field-Induced Decay of the Intraband Polarization 141
- 7.2.3 Wave Packet Revival in the Presence of Zener Tunneling 145

7.3 Summary 149
References 149

8 Electrical Transport Experiments 151
8.1 Modeling of Transport in Superlattices 151
- 8.1.1 Semiclassical Miniband Transport Theories 151
- 8.1.2 Scattering Mechanisms in Superlattices 154

8.2 Experimental Investigations of Miniband Transport 156
- 8.2.1 Investigations of Low-Field Transport 156
- 8.2.2 Resonances with Higher Minibands: Domain Formation 159

8.3 High-Field Transport: Observation of Negative Differential Conductivity 160
- 8.3.1 Experiments on Undoped Superlattices 161
- 8.3.2 Experiments on Doped Superlattices 162
- 8.3.3 Magnetotransport Experiments on Superlattices 162
- 8.3.4 AC Conductivity Measurements of Superlattice Transport 164
- 8.3.5 Superlattice Transport Measurements by Time-of-Flight Techniques 166

8.4 Details of Miniband Transport 169
- 8.4.1 Introduction 169
- 8.4.2 Coupling of Miniband Transport to the xy Dispersion . 169
- 8.4.3 Transport in Narrow Minibands: Suppression of Phonon Scattering and Influence of Disorder 171
- 8.4.4 Electrical Spectroscopy Using Transistor Structures ... 176

8.5 Negative Differential Conductivity in Bulk Silicon Carbide ... 179
8.6 Combined Transport and Optical Experiments 181
- 8.6.1 Comparison of Transport and Optical Experiments ... 181
- 8.6.2 Observation of Thermal Saturation in a Superlattice by Infrared Absorption 188

| | 8.6.3 | Electrical Transport Under Infrared Illumination ..... 190 |
| | 8.6.4 | Investigation of Zener Tunneling: Transport and Infrared Experiments .... 193 |

8.7 Summary... 196
References 196

9 Bloch Oscillations in Other Systems 201
9.1 Bloch Oscillations of Atoms............................... 201
9.2 Bloch Oscillations in Optical Wave Guides.................. 204
References 207

10 Future Prospects and Possible Applications................ 209
10.1 Basic Considerations for Coherent Electronic Devices 209
10.2 Device Concepts Using Coherent Transport Effects in Superlattices 211
    10.2.1 Detector Applications.............................. 211
    10.2.2 Free-Running Bloch Emitters 212
10.3 Bloch Lasers, or Is There Gain in a Biased Superlattice? 217
    10.3.1 Existence of Gain 218
    10.3.2 Stable Operation of a Bloch Oscillator 223
10.4 Comparison with Other Potential Sources of THz Radiation....................................... 227
10.5 Summary.. 229
References 229

Selected List of Symbols 233

Index 237

1 Introduction: Basics of High Field Transport

In this chapter, we first briefly discuss high-field transport in solids and introduce some of the key phenomena. Then, we provide a discussion of standard drift transport versus *ballistic* or *coherent* transport, the latter being typical in the case of Bloch oscillations. The key point is that the wave-like quantum-mechanical nature of the electrons becomes important when they reach the upper edge of the band. Instead of linear drift transport along the electric field direction, one obtains oscillations around a fixed point in space. These spatially localized oscillations compete with the effect of Zener tunneling, where the coupling to higher bands leads again to transport in the electric field direction.

1.1 Introducing Remarks

The transport of charged carriers in solids is a basic effect in physics and has far-reaching applications, which have considerable impact on our everyday life today. From a general standpoint, one might believe that the transport of a carrier in a solid consisting of approximately 10^{23} charged ions and conduction electrons, having many degrees of freedom, would be exceedingly difficult to describe. However, it can be shown that, after a series of simplifications, charge carrier transport can be described as a drift transport of quasi-free carriers with a velocity linearly proportional to the electric field. This linearity with respect to the field, which is generally known as *Ohm's law*, is valid in many practically relevant cases, such as current transport in metallic conductors.

The key simplifications required to obtain that result are (for the case of electrons)

- The electrons are close to a parabolic band minimum, where they can be described by a constant effective mass given by the reciprocal curvature of the band.
- The electrons stay in this extremum and are not transferred to another extremum.
- The electrons are subject to relaxation processes long before they reach a significant energy (compared with the width of the band they are moving in).

However, for very high fields, the simplifications made above start to fail, and Ohm's law is violated. One effect which has been known for a long time is the transfer of electrons to higher-lying minima with smaller curvature, i.e. larger effective mass (the Gunn effect [1]). The ensemble mobility then no longer linearly increases with the field, owing to the increasing portion of carriers which reach these higher minima. Finally, the carrier velocity saturates (e.g. in the case of silicon) or even decreases (e.g. for GaAs). In the latter case, the negative differential velocity leads to instabilities, which can be used for microwave generation (Gunn oscillator).

Another, even more general high-field transport effect occurs when the carriers gain enough energy to pass the inflection point of the band and reach the upper half of the band: the electron mass then becomes negative, and the carriers start to move *against* the electric field. If they are ballistically transported, i.e. not scattered at all, they will perform oscillations in space, but will not move on average. These *Bloch oscillations*, which are a direct consequence of the acceleration theorem found by Bloch in 1928, were first mentioned by Zener in 1934. It was long thought that this effect was a curiosity without practical consequences: typically, the band widths are so large that the carriers will be scattered long before they reach the upper parts of the band.

The situation was changed decisively by the introduction of semiconductor heterostructures. With these structures, *band gap engineering* became possible. The semiconductor superlattice invented by Esaki and Tsu [2] allowed one to realize artificial quasi–one–dimensional superlattices with small, adjustable band widths. In these structures, the carriers can reach the upper part of the band before being scattered. Accordingly, these structures turned out to be ideal to study the high-field transport effects which are discussed in this book.

1.2 Coherent vs. Incoherent Transport of Electrons in a Band

The standard picture of electronic transport in solids with a static electric field applied is that of drift transport, where, in the simplest case, the relation between the current density j and electric field F is given by Ohm's law,

$$j = \sigma F . \tag{1.1}$$

The underlying proportionality is between the drift velocity v_d and the electric field

$$v_\mathrm{d} = \mu F , \tag{1.2}$$

where μ is the mobility.[1]

---

[1] Note that, in general, both σ and μ are tensors. In our context, we can describe them as scalars since we usually deal only with transport perpendicular to the layers of a superlattice.

1.2 Coherent vs. Incoherent Transport

Drift transport was first considered in detail by Drude [3] (for a lucid discussion of Drude transport, see [4]): In this approach, carriers accelerate ballistically till they change their momentum in a scattering process. The drift velocity of the carriers is then determined by a balance between the momentum and energy gain from the field during the ballistic motion, and the momentum and energy changes owing to elastic and inelastic scattering processes.

In the Drude description, the conductivity σ of the ensemble average is given by

$$\sigma = \frac{e^2 n \tau}{m}, \tag{1.3}$$

where τ is the momentum relaxation time, n the carrier density, m the carrier mass, and e the elementary charge.

One of the central assumptions of the Drude picture is that the carriers move ballistically as quasi-free particles between scattering events. A description as free carriers with a constant (effective) mass is well justified in a solid if the carriers stay close to the band edges. This motion can then be fully described in a classical picture and corresponds for example to the downhill motion of a ball released on a tree-covered slope.

However, if the field is high enough that the carriers reach higher parts of the bands before they are scattered, the electrons in a solid do not behave like free electrons anymore: the effective mass (which is the inverse of the curvature of the band dispersion; see Fig. 1.1) starts to increase and becomes infinite at the inflection point of the band; above this point, the effective mass is negative, i.e. the carriers are decelerated by the field. These effects are caused by quantum mechanics and emphasize the importance of the wave-like

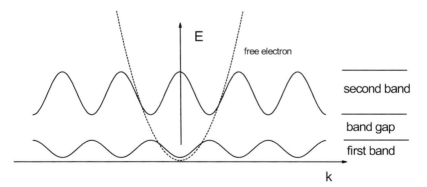

Fig. 1.1. Schematic energy dispersion of the first two bands in a periodic potential. The dispersion of the free electron is symbolized by the *dashed parabola*. For acceleration in a constant electric field (i.e., with constant k-space velocity), free carriers move up the parabola; in a solid without scattering, they follow the bands, going up and down in energy

nature of the carriers when transport processes are considered. The fact that the electron mass becomes negative is not a formal result, but indeed means that the direction of acceleration of the electron in Newton's law is changed: if the electron is first accelerated downwards in the potential when it has positive mass, it is accelerated upwards when the mass becomes negative. Clearly, the most striking effect of this quantum transport phenomenon is spatial oscillations of the carriers, the Bloch oscillations.

1.3 Basics of Bloch Oscillations

The most frequently used approach to discuss Bloch oscillations is a gedanken-experiment in a semiclassical one–dimensional picture. An electron is "put" at $k = 0$ (or, in more precise words, a wave packet with a narrow distribution in k-space around the bottom of the band at $k = 0$ is prepared). Then, an electric field F is instantaneously switched on. According to Bloch's theorem [5], the k-vectors of the electron distribution change in accordance with

$$\hbar \frac{\mathrm{d}k}{\mathrm{d}t} = eF \, , \tag{1.4}$$

i.e., the electron moves with constant velocity in k-space because of the electric field. Here, $\hbar$ is the Planck constant divided by 2π.

It is obvious from (1.4) that the k-vector of any state that the wave packet was initially composed of is given as a function of the time t by

$$k(t) = k_0 + eFt/\hbar \, , \tag{1.5}$$

where the field has been turned on at $t = 0$ and the initial wave vector was k_0. In other words, any given wave packet distribution in k-space will be linearly shifted as a whole with constant velocity in k-space.

Owing to the periodicity of the band structure (see Figs. 1.1 and 1.2), the electron moves thus periodically in energy along the dispersion of the band that it was put in [6]. Since the energy dispersion of a periodic solid is also periodic in k, the electron will gain energy until it is at the upper edge of the band. Then, it will lose energy till it reaches the center of the second Brillouin zone (Fig. 1.2). After crossing one Brillouin zone, the state of the electron is indistinguishable from the state it was in when it started its motion at $t = 0$ (except for a phase factor). The prediction of such periodic oscillations was actually first made by Zener in 1934 [7], so that the oscillations could also have been called Zener oscillations.

What is the real-space motion which corresponds to this oscillation in energy? For the simplest assumption of a harmonic dispersion of the band (which is obtained when the nearest–neighbor tight-binding model is applied; for details see Sect. 2.3.3), the dependence of the energy on the k-vector is given as

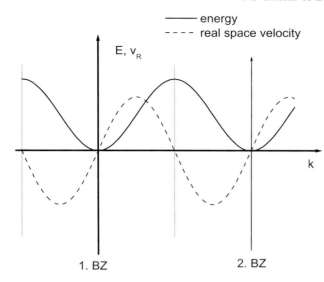

Fig. 1.2. Schematic drawing of the energy of a band (*solid line*) and associated real-space velocity (*dashed line*) as a function of the k-vector. The centers of the first and second Brillouin zones are labeled

$$E(k) = \frac{\Delta}{2} - \frac{\Delta}{2}\cos(kd) , \tag{1.6}$$

where Δ is the band width and d the periodicity of the structure. The real-space velocity v_R of an electron performing Bloch oscillations can be easily calculated:

$$v_R = \frac{1}{\hbar}\frac{\partial E}{\partial k} = \frac{d\Delta}{2\hbar}\sin(kd) , \tag{1.7}$$

i.e. the electron velocity depends harmonically on the k-vector and (owing to (1.5)) also on time. The electron thus moves back and forth in real space (see Fig. 1.3): in simple words, one can state that the electrons move till they hit the upper edge of the band and then return until they are again reflected at the lower edge of the band. Note that the total energy of the electrons (electronic energy in the solid plus field energy in the external potential) is constant: as we shall discuss below, Bloch oscillations are a coherent wave packet motion without energy exchange with an external system. The fact that the electron moves against the field when it has negative mass is thus not accompanied by a violation of energy conservation.

It is straightforward to derive the result that the spatial position of the electron also varies harmonically with time:

$$z(t) = z_0 - \frac{\Delta}{2eF}\cos(\omega_B t) \tag{1.8}$$

where

$$\omega_B = \frac{eFd}{\hbar} \tag{1.9}$$

and

$$T_{\rm B} = \frac{h}{eFd} = \frac{h}{E_{\rm BO}} \qquad (1.10)$$

are the angular frequency and time period, respectively, of the oscillation [7]. Here, z_0 is the initial spatial position of the electron wave packet and $E_{\rm BO}$ is the field energy from well to well (which is also the quantum energy of the Bloch oscillations), given by

$$E_{\rm BO} = eFd \,. \qquad (1.11)$$

The total spatial amplitude of the electron oscillation is the field-induced localization length

$$L = \frac{\Delta}{eF} \,. \qquad (1.12)$$

Note that the spatial extent and the period of the oscillations do not depend on the details of the band structure, but only on the band width and the electric field.

The Fourier transform of the Bloch oscillations in the time domain corresponds to a periodicity in the energy domain, which is associated with discrete, energetically equally spaced ladder states in a static bias field (see Chap. 2). The energies of the ladder states (which form the *Wannier–Stark ladder*) are given by

$$E_n = E_0 + n\,E_{\rm BO} \,, \qquad (1.13)$$

where the splitting between the ladder states $E_{\rm BO}$ is obviously identical with the electric–field energy from well to well.

It is obvious that both Bloch oscillations and the Wannier–Stark ladder are only observable if the carriers can travel through the full Brillouin zone before they are scattered, or, in other words, the scattering time τ (in (1.3)) needs to be longer than the Bloch oscillation period $T_{\rm B}$ (in (1.10)). To keep the Bloch oscillation period short, it is necessary to apply very high fields, which is not possible in metals owing to ohmic heating (see, e.g., [6]). In semiconductors, high fields can be applied more easily. Nevertheless, the high scattering rates make it very unlikely that the carriers can reach energies on the order of the band width, typically about 2 eV. Consequently, the observation of Bloch oscillations has not been reported for bulk materials.[2] There were some reports for the observation of a Wannier–Stark ladder in GaAs [8–11]. However, the observations did not unambiguously prove the existence of the Wannier–Stark ladder.

---

[2] One exception is the experiments on silicon carbide discussed in Sect. 8.5. However, SiC is a natural superlattice.

1.4 Coupling to Higher Bands: Zener Tunneling

The picture of Bloch oscillations just outlined is actually a strong simplification. The energy spectrum of a periodic structure does not contain one single band, but an infinite number of bands. Even for an arbitrarily small field, the lowest band is resonantly coupled to all higher bands. It is also obvious that all electrons will finally follow the field and will move downward in energy. In other words, no stationary states exist when a field is applied. In the following chapter 2, we shall briefly outline the long-standing theoretical debate about the electronic structure of a periodic potential in a static field when higher bands are not neglected.

We here briefly introduce the effect of Zener tunneling. Basically, we can distinguish three field regimes:

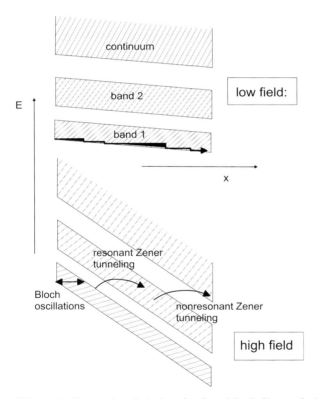

Fig. 1.3. Energy bands in low (*top*) and high (*bottom*) electric fields. For low fields, drift transport is dominant, for higher fields, carriers perform Bloch oscillations till they are transferred to higher bands by Zener tunneling. One can loosely distinguish a regime where tunneling to a state of one specific higher band is involved (resonant Zener tunneling), and one where a transition to a continuum resonance of many higher bands is dominant (nonresonant Zener tunneling)

- At low fields, when the carriers perform Drude transport, the coupling to higher bands is irrelevant (Fig. 1.3 top).
- For intermediate fields, the carriers start to perform Bloch oscillations. The coupling to higher bands is still weak and the dynamics of the Bloch oscillations are not influenced by this coupling. In Chaps. 3 to 5, we accordingly limit ourselves to the case where the electric field is so small that the field energy per period of the structure eFd is small compared with the potential barriers.
- For high fields (Fig. 1.3 bottom), there is also a probability that the carriers are transferred to a higher band by Zener tunneling. The theoretical debate mostly addressed the question of whether Zener tunneling is weak enough that Bloch oscillations are preserved. As we shall discuss later, this is, at least in semiconductor superlattices, clearly the case when the field is not too high.

In Chap. 7, we will discuss experiments where the field energy per period $E_{\rm BO} = eFd$ is comparable to the barriers and Zener tunneling becomes an important effect. However, as we shall discuss there, it needs a specifically designed structure with very low barriers to reach this regime in semiconductor superlattices.

References

1. J.B. Gunn, IBM J. Res. Dev. **8**, 141 (1964).
2. L. Esaki and R. Tsu, IBM J. Res. Dev. **14**, 61 (1970).
3. P. Drude, Ann. Phys. **1**, 566 (1900); Ann. Phys. **3**, 369 (1900).
4. N.W. Ashcroft and N.D. Mermin, *Solid State Physics,* (Saunders, Philadelphia, 1981), Chap. 1.
5. F. Bloch, Z. Phys. **52**, 555 (1928).
6. N.W. Ashcroft and N.D. Mermin, *Solid State Physics,* (Saunders, Philadelphia, 1981), p. 225.
7. C. Zener, Proc. R. Soc. London Ser. A **145**, 523 (1934).
8. A.G. Chynoweth, G.H. Wannier, R.A. Logan, and D.E. Thomas, Phys. Rev. Lett. **5**, 57 (1960).
9. A.G. Chynoweth, R.A. Logan, and D.E. Thomas, Phys. Rev. **125**, 877 (1962).
10. S. Maekawa, Phys. Rev. Lett. **24**, 1175 (1970).
11. R.W. Koss and L.M. Lambert, Phys. Rev. B **5**, 1479 (1972).

2 Semiconductor Superlattices

This chapter discusses the key properties of semiconductor superlattices, which are a class of artificial, new crystals created by the alternating deposition of different semiconductors. These superlattices allow one to "design" electronic properties such as band widths and dispersions over a large range by changing the geometrical parameters of the superlattice and the composition of the semiconductors it is composed of. By using this design freedom, it is possible to realize periodic structures which are much more suitable for observing Bloch oscillations than bulk semiconductors are.

After a brief introduction, we first describe the electronic properties of superlattices without an electric field. Then, we discuss the electronic properties under a static bias field, leading to the Wannier–Stark ladder. Finally, we briefly outline the optical properties and experimental observations of the Wannier–Stark ladder.

2.1 Introduction

In 1970, Esaki and Tsu proposed the semiconductor superlattice [1]: by means of the growth of alternating layers of semiconductors with smaller (in the well) and larger (in the barrier) band gaps (see Fig. 2.1), an artificial crystal is created in the growth direction.[1] The periodicity of this crystal in the growth direction is given by the sum $d = a + b$ of the thicknesses a and b of the small- and large-band-gap material, respectively. The layer of the small-band-gap material acts as a quantum well leading to quantum-confined states which can couple through the barriers formed by the larger band gap material. This leads to a formation of bands perpendicular to the layers called *minibands*. The eigenstates of these bands are delocalized Bloch-wave states (shown schematically in Fig. 2.1).

The key features of the semiconductor superlattice are that (i) the electronic structure can be controlled over a wide range by variation of the semiconductor materials, and (ii) the band widths can be made much smaller than

---
[1] We do not cover here the other type of superlattice, realized by the use of alternating layers of n-doped and p-doped material with the same band gap. For a review of the properties of these doping superlattices, see Ref. [2].

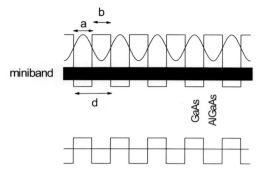

Fig. 2.1. Schematic picture of a semiconductor superlattice, made of semiconductors with smaller (e.g. GaAs) and larger (e.g. AlGaAs) band gaps. The periodicity d is given by the sum of well width a and barrier width b. The first electron and hole minibands and a miniband Bloch wave state of the electrons are also shown schematically

in conventional semiconductors, leading to a multitude of novel transport effects.

As we shall see later, the observation of such effects requires the controlled growth of tens or hundreds of layers with thicknesses of typically a few nanometers and interface roughnesses on the order of one monolayer. The growth techniques applied to achieve such structures are molecular-beam epitaxy (MBE) and metal-organic vapor phase epitaxy (MOVPE, often also referred as metal-organic chemical vapor deposition, MOCVD). These growth techniques have reached such a perfection that they have become, at least for the material systems frequently used, more or less an engineering discipline and routinely meet the requirements set out above. We therefore shall not further discuss growth techniques, but refer the reader to general reviews of the growth techniques of MBE [3] and MOVPE [4]. A detailed discussion of growth with respect to superlattices has been given by Fujiwara in [5].

The electronic structure of the superlattice will depend on the materials chosen and on the thicknesses of the layers. Each specific material combination leads to a particular relative arrangement of the valence bands and the conduction bands. We can distinguish three types (Fig. 2.2):

- In a *type I superlattice*, the bands of the smaller-band-gap material are sandwiched between those of the larger-gap material, forming potential wells for both electrons and holes in the low-band-gap material.
- In a *type II superlattice*, electrons have a potential well in one of the materials and holes have a potential well in the other.
- In the case of a *staggered-alignment superlattice*, the band gaps of the two materials do not overlap at all.

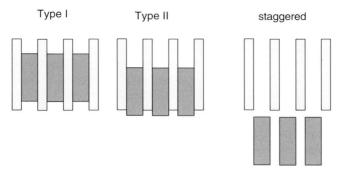

Fig. 2.2. Schematic picture of the different band-alignment types of semiconductor superlattices. The shaded regions represent the band gaps

All three types of superlattice offer interesting properties. Type I materials have been used most frequently, because they allow one to design narrow bands (minibands) for both electrons and holes with strong overlap of the wave functions, which have their maximum probability in the wells, thus leading to strong optical coupling. In type II materials, this overlap can be suppressed by transfer of the carriers to adjacent wells, leading to e.g. long carrier lifetimes even in direct semiconductors and strong fields owing to electron and hole separation, for example. In staggered-alignment materials, unusual transport properties can be achieved. In this book, we discuss only type I superlattices where both electrons and holes are confined in the same wells.

2.2 Electronic States in Superlattices Without an Electric Field

It is immediately obvious that the description of the electronic states of a semiconductor superlattice is a formidable task: if one were to try to calculate the properties from first principles, one would have to use a band structure calculation applied to a super-unit cell which contains the atomic layers of both the well and the barrier. Such calculations have been performed using the pseudopotential approach for single-monolayer superlattice layers [6] and for multilayer structures [7], and also using tight-binding methods [8]. However, most calculations of the electronic structures of superlattices have used various approximation methods.

The simplest approach is to assume a one-dimensional potential with constant energy in both the well and the barrier layers and an instantaneous potential jump between the two layers. This approach is identical to the Kronig–Penney model if the mass of a free electron is used. The Schrödinger equation for this potential can be readily solved, assuming that the wave functions and their derivatives are continuous at the interfaces (see, e.g., [9]); the solution yields a sequence of bands and band gaps. At least the lower-lying bands have nearly sinusoidal dispersion (1.6). Alternatively, the band structure can be obtained by the tight-binding approach, where the inclusion of only nearest-neighbor interactions leads to a harmonic dispersion.

In a standard semiconductor superlattice made by layer-by-layer growth, the full dispersion is three-dimensional since the carriers can move freely in x and y direction (Fig. 2.3). The (rather small) dispersion in the miniband direction is thus accompanied by a much larger dispersion in the other two directions. The total density of states of the superlattice is a superposition of the 1D dispersion of the minibands in z direction and the 2D dispersion in x and y directions, as shown in Fig. 2.4.

For a more precise calculation of the dispersion in a semiconductor superlattice, the calculation needs to include the effective masses of the carriers in

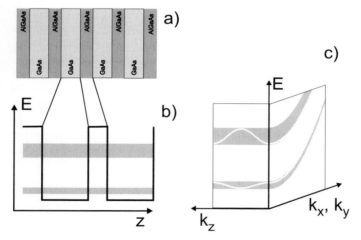

Fig. 2.3. Properties of a semiconductor superlattice in three dimensions: (**a**) structure (**b**) band edges (**c**) Dispersion in the z direction and in xy-directions

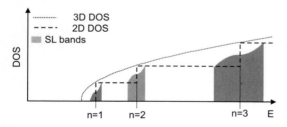

Fig. 2.4. Schematic illustration of the density of states for a bulk 3D system (*dotted parabola*), a 2D system (steps, *dashed line*), and a superlattice. The shaded areas indicate the minibands

2.2 Electronic States in Superlattices Without an Electric Field 13

the well and barrier, since they typically differ strongly from the free carrier mass. The Kronig–Penney model can be extended either by averaging the effective masses over the two materials or by using different masses with a suitable boundary condition for the derivative of the wave function. A precise description is further complicated by the fact that the electronic states which have to be considered are not at the band edge, but are above or below the edges owing to to the energy shifts caused by the quantization and the coupling. Thus, one needs to take the nonparabolicity, i.e. the dependence of the mass on energy, of the original bulk materials into account. It has been shown that the inclusion of such effects leads to deviations from the Kronig–Penney model. Figure 2.5 shows that the dispersion can deviate from a harmonic behavior.

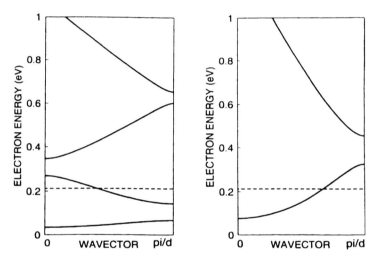

Fig. 2.5. k·p calculations of the dispersion for two GaAs/Al$_{0.3}$Ga$_{0.7}$As superlattices, with well/barrier widths of 7.0/2.0 nm (*left*) and 2.5/1.5 nm (*right*). The *dashed lines* mark the upper edge of the barriers (from [10])

A simple and conceptually elegant approach to describing the three-dimensional problem (which is applicable to semiconductor heterostructures in general) is the *envelope approximation* [11]. The basic approach is to separate the problem into the spatial dimensions parallel to the layers (denoted by x and y here, see Fig. 2.3) and the z direction (growth direction) perpendicular to the layers. The wave function is then written as a product of xy wave functions, which are assumed to be the Bloch waves of the bulk material, and a wave function in the z direction, which is assumed to be the Bloch wave of the bulk material multiplied by an envelope function which describes the influence of the heterostructure.

To give us a feeling for the dependence of the superlattice electronic bands on the structure, we present here some data computed by Bastard [11]. Figure 2.6 shows the bands and band gaps of an $Al_{0.3}Ga_{0.7}As$ superlattice structure with equal well and barrier widths $a = b$, i.e. $d = 2a$, as a function of the lattice periodicity d. For small periods, even the first miniband is not fully confined below the barrier edges (dashed line).

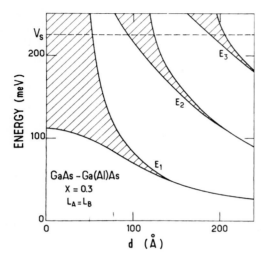

Fig. 2.6. Minibands and miniband gaps for an $Al_{0.3}Ga_{0.7}As$ superlattice with equal well and barrier widths as a function of the periodicity d. The *dashed line* marks the upper edge of the barrier (from [11])

Figure 2.7 shows similar data for $Al_{0.3}Ga_{0.7}As$ superlattice with well widths of 30 Å, 50 Å, and 70 Å when the barrier thickness is varied. It is obvious that the localized quantum well states start to couple in the lowest band at barrier widths of about 10 nm.

The band formation also depends on fluctuations in the growth parameters and phonon scattering, which both lead to broadening. If this broadening is larger than the miniband width, the formation of the miniband is suppressed. At low temperatures, the formation of a band starts at miniband widths of around 1 meV, when the broadening owing to imperfect growth is typically overcome. At room temperature, miniband widths of at least a few meV are required to have a truly delocalized system, owing to the strong phonon scattering.

The typical band widths of about 10–50 meV used for most of the experiments discussed in this book correspond to a barrier thickness of about 2–5 nm. An upper limit on the miniband width of the electrons is naturally given by the conduction band offset, which is in the GaAs/AlGaAs system, for example, about 0.5 eV at maximum.

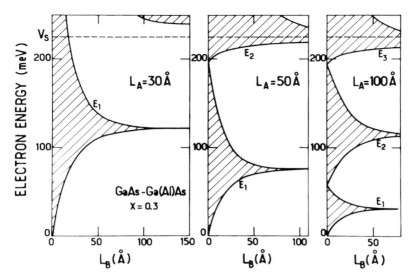

Fig. 2.7. Minibands and miniband gaps for $GaAs/Al_{0.3}Ga_{0.7}As$ superlattices with well widths a (here denoted by L_A) of 3, 5, and 7 nm as a function of barrier thickness b (here denoted by L_B). The *dashed lines* mark the upper edge of the barriers (from [11])

2.3 Electronic States in Superlattices with an Electric Field

2.3.1 Introduction

As discussed above, superlattices without an applied electric field are quasi-3D semiconductor structures. The eigenfunctions are Bloch waves which are spatially extended across the whole superlattice structure. If a static electric field is applied perpendicular to the layers of the superlattice, the electronic structure is markedly changed: the eigenstates of the periodic structure without a field, the well-known delocalized Bloch waves, are replaced by states which are partially localized.

Wannier [12] showed in 1960 that the energy spectrum of an electron in a periodic structure in an electric field F consists of equally spaced eigenvalues with energies $E_n = neFd$, where n is an integer and d is the lattice spacing.[2] The energy spacing eFd between the steps of the ladder can be understood in terms of the Fourier transform of the Bloch oscillations in time, as discussed above. Wannier also pointed out that the eigenfunctions cannot be normalized and should therefore not lead to stationary states.

---

[2] Note that an equidistantly spaced ladder of states had already been predicted in 1949 by James [13].

In contrast, some common approximations which were developed later such as the discrete model [14,15], the closely related tight-binding model [16, 17], and the one-band approximation [18,19], predict localized wave functions. Thus, it was a common belief that the spectrum of a Bloch electron in an electric field is discrete.

However, a discrete Wannier–Stark ladder can only be expected if only one band is taken into consideration [20]. Consequently, Avron et al. [21] showed that the energy spectrum of an electron in a biased periodic potential is in general *continuous* and confirmed that the electron states are not normalizable. However, despite the continuous spectrum, there are still equidistantly spaced resonances: from a practical point of view, the notion of "Wannier–Stark resonances" is justified when the broadening is much smaller than the level spacing [22]. Recently, Glück et al. [23] were able to analytically solve the Hamiltonian for a harmonic potential in a static field, thus analytically proving the existence of Wannier–Stark resonances. This work has been reviewed in [24].

The one-dimensional Hamiltonian for the envelope wave function of an electron in the z direction in a superlattice with an additional linear potential associated with an applied electric field F can be written as

$$\mathcal{H}\psi(z) = \left[-\frac{\hbar^2}{2m^*(z)} \frac{\partial^2}{\partial z^2} + V(z) + e\boldsymbol{F}z \right] \psi(z) = E\psi(z), \qquad (2.1)$$

where $V(z) = V(z+d)$ is the periodic potential of the superlattice.

This Hamiltonian can be solved analytically only under special conditions. Therefore, either numerical calculations are used to solve the eigenproblem (2.1), or one uses approximated electron wave functions utilizing the simplifications describe earlier.

Two interesting cases to look at are the limits of $V = 0$ and $F = 0$. For $V = 0$, the eigenproblem is equivalent to that of a triangular potential. In this case, the eigenfunctions are Airy functions, which have a plane wave tail on the low-energy side and decay exponentially in the barrier on the high-energy side. In case of a superlattice in a high field, the potential $V(z)$ in (2.1) becomes small compared with the external field and the electron wave function should resemble an Airy function. For $F = 0$, the Hamiltonian (2.1) consists of the kinetic energy, and the periodic potential of the superlattice. The translation symmetry leads to extended Bloch waves as previously discussed.

In the following, we discuss two approaches to solving the Schrödinger equation in the presence of an electric field: the Kane approach and the tight-binding approach.

2.3.2 The Kane Approach

The wave function of an electron in a biased superlattice can be expanded in a basis in real space or in $\boldsymbol{k}$-space. First, we describe the Kane model, which uses Bloch states as a basis.

2.3 Electronic States in Superlattices with an Electric Field

The Bloch functions

$$\varphi_{\boldsymbol{k}_z,\lambda}(\boldsymbol{z}) = u_{\boldsymbol{k}_z,\lambda}(\boldsymbol{z})\exp(\mathrm{i}\boldsymbol{k}_z\boldsymbol{z}) \qquad (2.2)$$

are a complete set of basis functions in the first Brillouin zone in $\boldsymbol{k}_z$ space where λ is the miniband index. The functions $u_{\boldsymbol{k}_z,\lambda}(\boldsymbol{z})$ represent the translation symmetry of the superlattice, with a periodicity d.

Therefore, the wave function $\psi(\boldsymbol{z})$ in the Schrödinger equation (2.1) can be expanded in terms of Bloch functions:

$$\psi(\boldsymbol{z}) = \sqrt{\frac{d}{2\pi}} \int_{-\pi/d}^{+\pi/d} \mathrm{d}\boldsymbol{k}_z \sum_\lambda \tilde{\psi}_\lambda(k_z)\varphi_{\boldsymbol{k}_z,\lambda}(\boldsymbol{z})\,, \qquad (2.3)$$

where $\tilde{\psi}_\lambda(\boldsymbol{k}_z)$ are the expansion coefficients in $\boldsymbol{k}_z$ space and d is the superlattice period. In the limit of an infinite superlattice, $\boldsymbol{k}_z$ becomes continuous. The electric field F is then treated as a perturbation. In the Kane model, the sum is restricted to a single band λ, and thus all intersubband interactions are neglected. The eigenfunctions $\psi_{\lambda,n}(\boldsymbol{z})$ for each band λ are known as Kane functions. The calculation leads to an eigenspectrum of equally spaced discrete energies

$$E_{\lambda,n} = E_{\lambda,0} + neFd\,, \qquad (2.4)$$

where $E_{\lambda,0}$ is the eigenenergy of an arbitrarily chosen central well, where $n = 0$. The energy difference between neighboring states is equal to the potential drop of the electric field eFd in one superlattice period.

Owing to the applied electric field, the translation symmetry breaks down and $\boldsymbol{k}_z$ cannot be taken as a good quantum number anymore. The new eigenstates are now labeled by the new quantum number n. The Kane model, as a one-band approximation, leads to localized wave functions which are normalizable and to a discrete energy spectrum. It is expected to result in good agreement with experiment for moderate electric fields but it fails in the high-field regime [25].

As mentioned above, an exact treatment of (2.1) leads to a continuous energy spectrum. This is not represented in the Kane model. Nevertheless, it is important to note that several eigenfunctions have a very low probability of tunneling towards the low-field side. A low tunneling rate corresponds to a high lifetime of the states, which stay localized in the wells. It was shown by Glutsch and Bechstedt [25], by calculating the electron states numerically, that certain eigenfunctions have a small tail towards the low-energy side and can therefore be taken as normalizable. These localized eigenfunctions are the wave functions the electronic wave packet is mainly composed of; they can be taken as quasi-stationary. This justifies approximations that lead to localized electron wave functions (which are normalizable) and to a discrete eigenspectrum.

2.3.3 The Tight-Binding Approach

In this section, wee describe the electron wave function in the framework of the tight-binding model. The tight-binding model is a real space approach which has been used by various authors, e.g. Bastard et al. [11, 16] and Dignam et al. [17].

The general form of the eigenfunction of a carrier in the one-band tight-binding approach can be written as

$$\psi_{n,\lambda}(z) = \sum_m C_n(m)\chi_{\text{loc},\lambda}(z - md), \tag{2.5}$$

where n is the index of the eigenstate, λ is the band index, and the $\chi_{\text{loc},\lambda}$ are the localized wave functions of electrons or holes in a single-well potential. The envelope function $C_n(m)$ of a superlattice for zero field is a plane wave that maintains the translation symmetry. As already mentioned above, Wannier [12] showed that the additional electric field changes the solutions of the Hamiltonian entirely: the quasi-continuous energy states of the miniband split, separated by the so-called Stark spacing eFd, which equals the potential drop between two adjacent wells. Such a set of quantized energies for each miniband λ forms again a Wannier–Stark ladder

$$E_{\lambda,n} = E_{\lambda,0} + neFd, \tag{2.6}$$

where $E_{\lambda,0}$ is the discrete eigenenergy of the central well ($z = 0$). This is exactly the same result as was found in the Kane approximation.

The envelope functions $C_n(m)$ are modified in a biased superlattice. The resulting Wannier–Stark ladder eigenfunctions in the nearest-neighbor approximation (ignoring wave function overlap, even between adjacent wells) can be written as [16, 26]

$$\psi_{n,\lambda}(z) = \sum_{m=-\infty}^{+\infty} J_{m-n}(\theta)\chi_{\text{loc},\lambda}(z - md), \tag{2.7}$$

where $J_n(\theta)$ is a Bessel function of the first kind. The index n defines the well on which the wave function is centered: the n th wave function is centered on the n th well. The translation symmetry of the eigenfunctions is represented by the relation

$$\psi_{n,\lambda} = \psi_{0,\lambda}(z + nd). \tag{2.8}$$

The spatial extension of a Wannier–Stark ladder eigenfunction over adjacent wells is strongly dependent on the electric-field strength (see Fig. 2.8). A closely related phenomenon is the localization of the carrier wave function with increasing electric field. In the zero-field case ($F = 0$), the superlattice is a system of degenerate quantum wells. Therefore, the electron is spread over the whole superlattice in this case.

An increased electric field leads to eigenstates which are shifted more and more out of resonance with the states in the neighboring wells. The wave

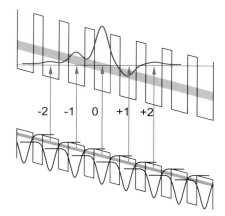

Fig. 2.8. Schematic drawing of tight-binding electron and hole eigenfunctions ($n = 0$) for a medium field. The electron wave function is already quite localized but still has a probability density in adjacent wells. This provides the possibility of interband transitions from hole states, which are localized in different wells (indicated by *arrows*). Further localization of the electron wave function with increasing field clearly causes the nonvertical Stark transitions to vanish. The *shaded areas* indicate the tilted 1D minibands

function tends towards a single-well wave function, and the system ultimately resembles the quasi-2D quantum structure of a quantum well. This *field-induced localization* is described by the parameter θ in the Bessel function, which is the ratio of the classical localization length L to the superlattice period d:

$$\theta = \frac{L}{d} = \frac{\Delta}{eFd}. \tag{2.9}$$

For a medium field, where $\theta \ll 1$, the sum (2.7) is dominated by one term only, the Bessel function of zero order ($m - n = 0$). The parameter θ describes the suppression of contributions that extend the wave function to neighboring wells, because $J_n(\theta)$ for $n > 0$ goes to zero as $\theta \to 0$ with increasing field. The field strength at which the wave function localizes scales inversely with the miniband width Δ. At high fields, the wave function in the tight-binding approximation approaches the single-well function. From another point of view, field-induced localization arises from the impossibility of achieving resonant tunneling over the whole superlattice if there are no degenerate states [26]. It is then obvious that the localization breaks down when tunneling to other subbands is possible (see Chap. 6).

2.4 Optical Transitions in Biased Superlattices

2.4.1 General Picture

We now discuss optical interband transitions from hole states in the valence band to electron states in the conduction band. Each miniband E_λ^e in the conduction band and E_λ^h in the valence band gives rise to a Wannier–Stark-ladder. Owing to the large mass of heavy holes, their coupling to neighboring wells is very small and they have a very small miniband width. Thus, even at small fields, the heavy holes can be considered as localized in one well. Therefore, optical interband transitions occur preferentially between a localized heavy hole and an extended electron wave function, as schematically shown in Fig. 2.8. Since the system is translationally symmetric in the z direction, the transition rate for each absorption line of one example of a well just needs to be multiplied by the number of wells.

We use the Wannier–Stark ladder index n and the miniband index λ to label the various possible Wannier–Stark ladder transitions. We also need to distinguish between heavy-hole and light-hole states. Figure 2.9 shows how the Wannier–Stark ladder transitions evolve with the electric-field strength. The spectral separation of two Wannier–Stark ladder transitions increases linearly with the field. The plot of Wannier–Stark ladder transitions as a function of electric field resembles a fan and is, accordingly, called a "fan chart".

Transitions from a hole localized in a given well to an electron in the same well are called *vertical transitions*; transitions from a hole localized in

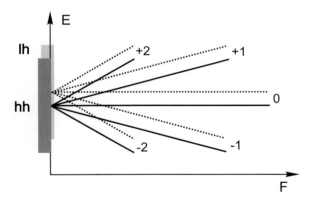

Fig. 2.9. Schematic drawing of the absorption peak positions vs. field. The Wannier–Stark ladder transitions evolve from the center of the 1D miniband and can be resolved when the field splitting is larger than the line width. The resonance frequency of each Wannier–Stark ladder transition increases linearly with the field according to $neFd$. Transitions originating from heavy-hole (*solid lines*) and light-hole (*dotted lines*) states are shown. Owing to a higher quantization energy of the light hole in the valence band well, its miniband and the associated Wannier–Stark ladder are shifted to higher energies

a given well to an electron centered on a neighboring well ($n \neq n'$) are called nonvertical Wannier–Stark ladder transitions. These interband transitions are labeled $hh_{\pm n}$ and $lh_{\pm n}$ for heavy-hole and light-hole transitions, respectively. If no additional index is used, transitions from the first miniband in the valence band to the first miniband in the conduction band are meant here ((λ, λ') = (1, 1)).

The oscillator strength of a transition, apart from depending on the density of states, depends strongly on the spatial overlap of the electron and hole wave functions of the transition considered. With increasing electric field, the field-induced localization leads to a decreasing extension of the wave function. Thus, if the resonance frequency of a nonvertical Wannier–Stark ladder transition ($n > 0$) shifts spectrally much further away than the miniband width at a certain electric field, its oscillator strength will vanish. Each Wannier–Stark ladder state is associated with a subband accommodating states of free in-plane motion. For medium fields, the density of states resembles the two-dimensional case, for medium fields. For fields at which the electron is mainly localized in the center well, mainly the hh_0 transition contributes to the absorption.

The selection rules for the Wannier–Stark ladder transitions, which are defined by the envelope function in the z direction, can be derived from the single-well wave functions associated with the eigenstate λ. Transitions between minibands with envelope wave functions of different parity are forbidden. An applied electric field breaks the symmetry, and therefore strongly forbidden transitions become weakly allowed. This phenomenon becomes visible at higher fields.[3]

2.4.2 The Excitonic Wannier–Stark Ladder

It is important to point out that the optical experiments used for tracing the Wannier–Stark ladder structure do not probe a single-particle ladder in the electronic system, but rather excitonic states. Therefore, a more precise description of the absorption energies and spectra of a superlattice requires a two-particle picture. The two-particle Hamiltonian can be separated into the in-plane motion describing 2D excitons and the motion in the z direction.

This problem was first considered by Dignam and Sipe [27]. They treated the optical spectrum of a biased semiconductor superlattice and showed that the Coulomb interaction leads to significant changes. Figure 2.10 shows an example of the results of their calculations.

The effects of the excitonic interaction can be summarized as follows:

- The Wannier–Stark ladder transitions can be resolved as excitonic peaks in the linear absorption. These transitions are accompanied by a correspond-

---

[3] Note that for a detailed description, the symmetry properties of the underlying Bloch waves of the bulk semiconductors must be considered as well.

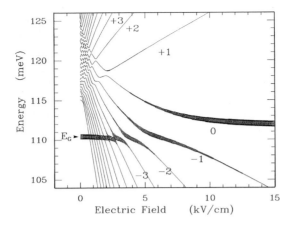

Fig. 2.10. Theoretically calculated fan chart of the transitions in a superlattice, including excitonic interactions. The width of the lines symbolizes the oscillator strength (from [27])

ing exciton continuum which is observed as a small step separated by the exciton binding energy.
- In general, excitonic effects redistribute oscillator strength from transitions at higher energy to those at lower energy owing to the attractive nature of the Coulomb potential. In an unbiased superlattice, the absorption is dominated by a single exciton peak due to the first bound exciton state, and a plateau rising at the low-energy side of the single-particle miniband. For a biased superlattice, each Wannier–Stark ladder transition is associated with a strong excitonic peak. The transitions to bound excitonic states gain considerably oscillator strength compared with the single-particle absorption.
- A strong asymmetry between positive and negative Wannier–Stark ladder transitions with the same index n is observed. The transition to the low-energy side (($-n$) transition) gains a higher oscillator strength than does the ($+n$) Wannier–Stark ladder transition. The Coulomb potential superimposed on the electric field results in a modified effective potential. Electrons excited into a ($-n$) Wannier–Stark ladder transition have a higher probability density at the central well ($z = 0$) than do electrons which have undergone ($+n$) Wannier–Stark ladder transitions. This results in a higher oscillator strength for the ($-n$) transition. This asymmetric localization behavior is also expected to have an impact on the exciton binding energy, as shown by Dignam and Sipe [27]. For medium fields, the ($-n$) transition has a considerably higher exciton binding energy than does its positive counterpart denoted by ($+n$). With increasing field, the difference vanishes because the Coulomb potential can be neglected in the high-field limit. The direct ($n = 0$) Wannier–Stark ladder transition clearly has the highest exciton binding energy.
- The transitions of the Wannier–Stark ladder consequently become nonequidistant owing to the different excitonic binding energies of the Wan-

nier–Stark ladder transitions. The largest binding energy occurs for the $n = 0$ transition owing to its large spatial overlap; indirect transitions have correspondingly lower binding energies.
• The Coulomb coupling of bound excitonic states and degenerate continuum states of subbands of lower Wannier–Stark ladder transitions leads to Fano resonances, which have been investigated experimentally in superlattices [28]. Fano resonances, in absorption, lead to a typical asymmetric line shape of resonant transitions. The line has a fast-rising side with a dip under the continuum absorption level, and a slow-rising low-energy edge. It was also shown that the coupling influences the line width and consequently the dephasing behavior of the excitonic system [28].

We shall discuss the influence of excitonic effects on the dynamics of Bloch oscillations in Sect. 4.2.

2.4.3 Experimental Observation of the Wannier–Stark Ladder

In this section, we briefly discuss experiments which have addressed the linear optical properties of superlattices. The typical techniques are photocurrent spectroscopy and linear absorption.

As mentioned above, there was a long-standing debate about the electronic structure of a periodic potential with a linear electric field superimposed. With the advent of the semiconductor superlattice, it became possible to address this question experimentally. The breakthrough came with optical studies of biased superlattices: the theoretical debate about the existence of the Wannier–Stark ladder was conclusively settled by its experimental observation, which directly showed the existence of ladder-like states forming a fan chart when the electric field was varied.

The Wannier–Stark ladder was first observed experimentally in GaAs/AlGaAs superlattices in 1988 by Mendez et al. [29] and Voisin et al. [30]. The former experiment demonstrated the existence of the resonances by photocurrent spectroscopy of a biased superlattice, and the latter investigated the absorption of a biased superlattice.

Figure 2.11 shows photocurrent spectra for a 30 Å GaAs/ 35 Å $Al_{0.35}Ga_{0.65}As$ superlattice taken at low temperature. Owing to the existence of excitons, the transitions of the Wannier–Stark ladder are nicely visible as field-dependent peaks. Both heavy-hole and somewhat weaker light-hole transitions are visible.

The corresponding fan chart is displayed in Fig. 2.12 and clearly shows the fan-like electronic structure of the system, for both the heavy-hole and the light-hole transitions. In the other experiment [30], the Wannier–Stark ladder was observed in the absorption spectrum.

The Wannier–Stark ladder has been investigated by a multitude of experiments since then. For instance, the spatial extension of the coherence of the wave functions was investigated by tracing the nonvertical transitions as a function of temperature [31].

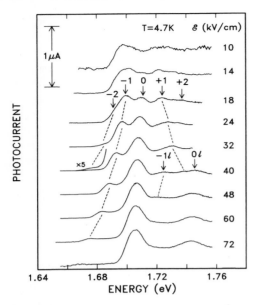

Fig. 2.11. Photocurrent spectra of a 30 Å GaAs/ 35 Å Al$_{0.35}$Ga$_{0.65}$As superlattice as a function of the electric field. The heavy-hole transitions are labeled by numbers from −2 to +2; the light-hole transitions are labeled −1l and 0l (from [29])

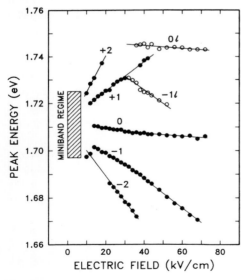

Fig. 2.12. Fan chart of transitions in the photocurrent spectra, showing the ladder-like electronic structure (from [29])

2.5 Summary

In this chapter, we have discussed some basic properties of semiconductor superlattices. The brilliant idea of Esaki and Tsu of realizing an artificial crystal with tunable properties has stimulated a large amount of theoretical and experimental research. In particular, semiconductor superlattices have helped to resolve some questions about the electronic states in periodic potentials which had been controversial for a long time.

References

1. L. Esaki and R. Tsu, IBM J. Res. Dev. **14**, 61 (1970).
2. G.H. Doehler, IEEE J. Quantum Electron. QE **22**, 1682 (1986); G. Doehler "Doping Superlattices - Historical Overview", in *Properties of III–V Superlattices and Quantum Wells*, ed. P.K. Bhattacharya, Data Review Series No. 15 (Electronic Materials Information Service (EMIS) of the Institution of Electrical Engineers, London 1996), p. 11.
3. E.H.C. Parker, ed., *The Physics and Technology of MBE* (Plenum, New York 1985).
4. G.B. Stringfellow, *Organometallic Vapor-Phase Epitaxy: Theory and Practice*, 2nd ed. (Academic Press, New York 1998).
5. K. Fujiwara, in *Semiconductor Superlattices – Growth and Electronic Properties*, ed. H.T. Grahn (World Scientific, Singapore 1995), p. 1.
6. E. Caruthers and P.J. Lin-Chang, Phys. Rev. Lett. **38**, 1543 (1977).
7. M.A. Jaros, K.B. Wong, and M.A. Gell, Phys. Rev. B **31**, 1205 (1975).
8. J.N. Schulman and T.C. McGill, Phys. Rev. Lett. **39**, 1680 (1977).
9. K. Seeger, *Semiconductor Physics*, 8th ed. (Springer, Berlin Heidelberg 2002).
10. A. Sibille, in *Semiconductor Superlattices - Growth and Electronic Properties*, ed. H.T. Grahn (World Scientific, Singapore 1995), p. 29.
11. G. Bastard, *Wave Mechanics Applied to Semiconductor Heterostructures* (Les Editions de physique, Les Ulis 1988), p. 63.
12. G.H. Wannier, Phys. Rev. **117**, 432 (1960).
13. H.M. James, Phys. Rev. **76**, 1611 (1949).
14. K. Hacker and G. Obermair, Z. Phys. **234**, 1 (1970).
15. M. Luban and J. H. Luscombe, Phys. Rev. B **35**, 9045 (1987).
16. J. Bleuse, G. Bastard, and P. Voisin, Phys. Rev. Lett. **60**, 220 (1988).
17. M. Dignam, J.E. Sipe, and J. Shah, Phys. Rev. B **49**, 10502 (1994).
18. E.O. Kane, J. Phys. Chem. Solids **12**, 181 (1959).
19. J. Callaway, *Quantum Physics of the Solid State*, Part B (Academic Press, New York 1974), pp. 465.
20. J. Zak, Phys. Rev. Lett. **20**, 1477 (1968).
21. J.E. Avron, J. Zak, A. Grossmann, and L. Gunther, J. Math. Phys. **18**, 918 (1977).
22. A. Nenciu and G. Nenciu, J. Phys. A **15**, 3313 (1982).
23. M. Glück, A.R. Kolovsky, H.J. Korsch, and N. Moiseyev, Eur. Phys. J. D **4**, 239 (1998).
24. M. Glück, A.R. Kolovsky, and H.J. Korsch, Phys. Rep. **366**, 103 (2002).

25. S. Glutsch and F. Bechstedt, Phys. Rev. B **57**, 11887 (1998).
26. G. Bastard, J. Bleuse, R. Ferreira, and P. Voisin, Superlatt. Microstruct. **6**, 77 (1989).
27. M.M. Dignam and J.E. Sipe, Phys. Rev. Lett. **64**, 1797 (1991).
28. C.P. Holfeld, F. Löser, M. Sudzius, K. Leo, D.M. Whittaker, and K. Köhler, Phys. Rev. Lett. **80**, 874 (1998).
29. E.E. Mendez, F. Agullo-Rueda, and J.M. Hong, Phys. Rev. Lett. **60**, 2426 (1988).
30. P. Voisin, J. Bleuse, C. Bouche, S. Gaillard, C. Alibert, and A. Regreny, Phys. Rev. Lett. **61**, 1639 (1988).
31. E.E. Mendez, F. Agullo-Rueda, and J.M. Hong, Appl. Phys. Lett. **56**, 2545 (1990).

3 Interband Optical Experiments on Bloch Oscillations: Basic Observations

Up to now, most experiments which have investigated Bloch oscillations in superlattices have employed optical methods to excite and/or detect the oscillatory motion. The simple reason behind that is that optical generation with a short-pulse laser produces a carrier ensemble with a uniform phase. Thus, the motion of this ensemble in k-space is synchronized for all carriers and leads, for example, to a macroscopic dipole moment oscillating with the Bloch frequency, which can be directly detected. In contrast, such a synchronized motion is not easily achievable in transport measurements. Experiments with optical excitation of Bloch oscillations have been reviewed in [1–4].

How the optical experiments are related to transport experiments is an interesting question. In this chapter, we therefore first discuss the nature of the optical experiments and their relation to transport experiments. Then, we address the basic interband optical experiments which have been performed to detect Bloch oscillations. Experiments which address the spatial dynamics using optical techniques are outlined in Chap. 4; experiments where THz radiation is detected are discussed in Chap. 5.

3.1 Coherent Optical Excitation of Superlattices: Relation to the Gedankenexperiment

There is no immediate connection between transport and optical experiments in semiconductors. Therefore, it is necessary to briefly discuss the nature of the two types of experiments and their relation. We follow closely a description given in [5].

In the optical experiments, interband excitation leads to the generation of electrons and holes. The distribution of the optically generated carriers depends sensitively on the conditions of the excitation. An excitation with a pulsed laser source with a pulse duration shorter than the Bloch oscillation period will generate wave packets with a distribution in k-space which is related to the pulse spectrum of the laser.

One difference between all optical experiments performed so far and the gedankenexperiment outlined in Sect. 1.3 is the fact that the static electric field is applied permanently, and not switched on after excitation. It is thus easier to discuss these experiments in the context of the Wannier–Stark ladder, i.e. when a superposition of the quasi-eigenstates of the biased superlattice is considered.

Short-pulse laser excitation of the Wannier–Stark ladder states will lead to an electronic wave packet of those states coupled via a common hole state. Because the spacing of the ladder is given by eFd, it is quite obvious that the wave packet will perform oscillations (quantum beats) with an angular frequency given by

$$\omega_B = \frac{eFd}{\hbar}, \tag{3.1}$$

i.e. the period of Bloch oscillations. However, there are potentially two differences from the gedankenexperiment which need to be discussed in detail:

- It is not a priori obvious how this wave packet, composed of Wannier–Stark ladder states, is related to the wave packet in k-space described in Sect. 1.3.
- The Coulomb interaction between electrons and holes will lead to a modification of the electronic states which does not exist in the transport experiments, where only one type of carrier is present.

One has thus to answer the question of whether the optical experiments observe quantum beats of the Wannier–Stark ladder states are related to that Bloch oscillations. In the following, we show that the quantum beats and Bloch oscillations are the same physical phenomenon, although with some small modifications.

The argument is as follows [6, 7]. The electron which, in the gedankenexperiment, is "put" at $k = 0$ is actually a wave packet of Bloch wave states (which are the eigenstates of the system *without* a field) localized around $k = 0$. If the static field is turned on, the k-states are no longer eigenstates; they become time-dependent following (1.5). This time dependence causes an oscillatory motion owing to the periodicity of the energy bands.

The Bloch wave eigenstates form a complete basis of eigenstates in the absence of a field. Similarly, the Wannier–Stark states form a complete basis in the case with an applied field. The wave packet can thus be expressed also as a sum of Wannier–Stark states (the eigenstates *with* an electric field) with certain weights. The two representations are linked by a unitary transformation. For the optical experiments, it is easier to imagine the photogenerated wave packet as a superposition of Wannier–Stark states.

The key point when comparing the experiments is whether the spatial distribution and the temporal and spatial dynamics of an optically generated wave packet can be made identical to the k-space wave packet of the gedankenexperiment. It turns out that this is possible with a suitable choice of the wave packet parameters, which can be controlled by the spectrum of the optical excitation. The only difference between the gedankenexperiment and the optical experiments is the fact that optical experiments at the interband frequency are related to the photoexcitation of holes, which influence the electronic properties of the system. As we shall discuss below, this leads to some changes in the eigenstates and their dynamics. However, for fields that are not too low, the differences are rather small.

The spatial dynamics of photogenerated wave packets have been investigated theoretically for Bloch oscillations in semiconductor superlattices [8,9]. The spatial amplitude of the Bloch wave packet oscillation depends strongly on the optical excitation conditions. For most cases, however, the center of mass of the wave packet performs a harmonic spatial motion.

The most detailed theoretical study including the influence of excitons was performed by Dignam et al. [8]. These authors showed that the spatial amplitude of the Bloch-oscillating wave packet can reach the semiclassical value given by (1.12) for certain parameters of the laser excitation, if excitonic effects are neglected. The inclusion of excitonic effects leads to some small changes (not surprisingly, mainly a reduction of the amplitudes), but more than 90% of the spatial amplitude of the gedankenexperiment is reached. It can thus be concluded that the optical experiments can reproduce the dynamics of the transport gedankenexperiment almost completely. In Sect. 4.1, we discuss the fact that the spatial displacement of optically excited wave packets can be measured experimentally and agrees well with the theoretical expectations.

3.2 Interband vs. Intraband Dynamics in Optical Experiments

3.2.1 Introduction

In the Wannier–Stark ladder picture, it is obvious that the dynamics of the Bloch-oscillating electrons are caused by the relative temporal development of the eigenstates that the wave packet is composed of. The total wave function is a superposition of electronic eigenstates ψ_i (having energies E_i) with certain weights a_i. The laser-generated wave packet Φ is then a quantum mechanical superposition of eigenstates ψ_i of the form

$$\Phi = \sum_j a_j \psi_j e^{iE_j t/\hbar} \ . \tag{3.2}$$

It is obvious that the probability $|\Phi|^2$ will show oscillations with beat frequencies given by the energy differences $E_i - E_j$ between the various eigenstates that the nonstationary superposition is composed of. The dynamics will only be present as long as there is coherence between the different parts of the wave packet, i.e. the different parts have well-defined phases. Bloch oscillations are thus characterized by *intraband coherence*, i.e. coherence of the states of the conduction band.

We emphasize here that this intraband coherence has to be distinguished from the *interband coherence* between the valence and the conduction band. Such a coherence, which is relevant in the case of a superposition of a hole and an electron state, can be created by excitation with coherent light, for example. In the optically excited Bloch oscillation experiments, a short, spectrally broad laser pulse creates an *interband* coherence between a valence ground state and a number of excited states (the Wannier–Stark ladder states). The generation of the interband coherence with a short, coherent light pulse is then also associated with well-defined phase relations between the excited states, or, in other words, an *intraband* coherence is set up as well. In the following, we discuss the optical excitation processes for a two-level system first and then extend our treatment to a three-level system, which is the simplest case in which Bloch oscillations are possible.

The simplest way to model an optical transition from the valence to the conduction band in a semiconductor is to consider an ensemble of two-level systems, described in the density matrix formalism. The light field creates a superposition of the ground and excited state

$$\psi = c_1 \psi_g + c_2 \psi_e , \tag{3.3}$$

where the occupation oscillates between the ground and the excited state at the Rabi frequency as long as the excitation electric field is turned on. The density matrix for the two-level system reads

$$\rho = \begin{bmatrix} \rho_{gg} & \rho_{ge} \\ \rho_{eg} & \rho_{ee} \end{bmatrix} . \tag{3.4}$$

The diagonal states of the density matrix describe the population in the ground state g and in the excited state e, respectively. The off-diagonal density matrix elements describe the coherence between the ground and the excited state induced by the laser, which determines the total polarization of the system. These elements are only nonzero if the relative phase of the ensemble of excited states is preserved, i.e. if the coherence obtained from the exciting laser light is still present. The microscopic dipole moment of each individual two-level system then contributes to a macroscopic dipole moment of first order, which is proportional to the off-diagonal elements:

$$P^{(1)}(t) \propto \rho_{eg} . \tag{3.5}$$

Scattering processes will lead to a buildup of population in the excited state proportional to the difference between the diagonal matrix elements of the density matrix:

$$n(t) = \rho_{ee} - \rho_{gg} . \tag{3.6}$$

The temporal evolution of the density matrix is described by a Liouville equation which describes the temporal variations of the density matrix elements caused by driving terms related to the field and by scattering terms. This equation can be written as

3.2 Interband vs. Intraband Dynamics in Optical Experiments

$$\frac{\partial}{\partial t}\rho_{\text{eg}} = \frac{\partial}{\partial t}\rho_{\text{eg}}|_{\text{coh}} + \frac{\partial}{\partial t}\rho_{\text{eg}}|_{\text{scatt}} \tag{3.7}$$

for the terms relevant to the polarization and

$$\frac{\partial}{\partial t}\rho_{\text{ee}} = \frac{\partial}{\partial t}\rho_{\text{ee}}|_{\text{coh}} + \frac{\partial}{\partial t}\rho_{\text{ee}}|_{\text{scatt}} \tag{3.8}$$

for the population terms. The coherent part is driven by the optical excitation:

$$\frac{\partial}{\partial t}\rho_{\text{eg}}|_{\text{coh}} = (H_{\text{ge}}\rho_{\text{eg}} - H_{\text{eg}}\rho_{\text{ge}}) \; , \tag{3.9}$$

where

$$H_{\text{eg}} = H_{\text{ge}}^{*} = -\mu_{\text{eg}}\tilde{E} \; . \tag{3.10}$$

Here μ_{eg} is the dipole moment of the transition and $\tilde{E}$ is the laser field. These equations are the optical Bloch equations, which are an equivalent of the Bloch equations for a magnetic two-level system.

A microscopic description of scattering processes in semiconductors is quite involved. Usually, therefore the scattering terms are approximated by a relaxation time ansatz

$$\frac{\partial}{\partial t}\rho_{\text{eg}}|_{\text{scatt}} = -\rho_{\text{eg}}/T_2 \tag{3.11}$$

and

$$\frac{\partial}{\partial t}\rho_{\text{ee}}|_{\text{scatt}} = -\rho_{\text{ee}}/T_1 \; , \tag{3.12}$$

where the time constants T_2 and T_1 are the transverse and longitudinal relaxation times, respectively.

If the polarization is generated by a short laser pulse, it will radiate freely after the pulse is over. In a real structure, the polarization will also decay owing to dephasing processes. There are several different possibilities which can lead to a decay:

- *First*, inhomogeneous dephasing will occur if the two-level systems do not have all the same energy, but are distributed in energy, for example because of crystal imperfections. With increasing time, the excitations will get out of phase, and the macroscopic polarization will completely disappear if the polarization vectors become randomly distributed in phase. Since the interference is not caused by a statistical process, the excitation can be rephased, for instance in nonlinear optical techniques, as discussed below in Sect. 3.3.1.
- *Second*, besides the polarization decay due to inhomogeneous broadening, a decay due to scattering will occur. For an ensemble of two-level systems with no inhomogeneous broadening, the decay of the first-order polarization is exponential:

$$P^{(1)}(t) = P^{(1)}(0) \times e^{-t/T_2} \; . \tag{3.13}$$

This decay corresponds to pure homogeneous dephasing with a dephasing time T_2. The homogeneous line width of the transition due to scattering processes is then given by the scattering rate γ, which is related to the time constant of the homogeneous first-order polarization decay by

$$\gamma = \frac{h}{\pi T_2} \tag{3.14}$$

or, in commonly used units,

$$\gamma \, [\text{meV}] = \frac{1.31}{T_2 \, [\text{ps}]}. \tag{3.15}$$

- *Third*, the return to the ground state due to recombination (with a time constant T_1) will reduce the macroscopic polarization. Since the timescale of the recombination process is typically on the nanosecond scale, this process is not important for Bloch oscillations, which have damping times on the picosecond timescale.

3.2.2 Description of Bloch Oscillations

The simplest approach to modeling Bloch oscillations is to extend the two-level system to a three-level system with a ground state $|g\rangle$ and two excited states $|e1\rangle$ and $|e2\rangle$, as the first step. In this description, the density matrix contains, besides the three population terms on the diagonal, six (2×3) nondiagonal elements which describe the coherence between the different transitions. The three independent matrix elements refer to the two *interband* coherences from the ground state $|g\rangle$ to the two excited states $|e1\rangle$ and $|e2\rangle$, related to damping rates $\gamma_{g,e1}$ and $\gamma_{g,e2}$, respectively, and to the *intraband* coherence between the two excited states $|e1\rangle$ and $|e2\rangle$, related to a damping rate $\gamma_{e1,e2}$. This intraband coherence is the relevant coherence for the existence of Bloch oscillations.

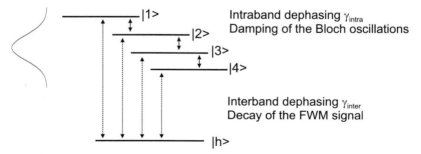

Fig. 3.1. Schematic illustration of the various transitions and coherences in the Wannier–Stark ladder. The *dotted arrows* mark the interband coherence, and the *solid arrows* the intraband coherence. The excitation of a wave packet by a spectrally broad laser pulse is indicated at the left side

The three-level system can then be easily extended to a multilevel system, which is the simplest description of a biased superlattice. Here, every hole state can be coupled to a number of electron states which have a significant spatial overlap with the hole state. In a single-particle model, these different excited states all have the same probability distribution, but are characterized by different translations relative to a reference well. Thus, one could define a common damping time of the coherence between hole states and electron states γ_{inter}. The coherence relevant to the Bloch oscillations is then the coherence between the electron states, with which the damping time γ_{intra} is associated. The situation is schematically described in Fig. 3.1.

3.2.3 Extension to Real Semiconductors

An optical excitation in a real semiconductor clearly cannot be described by a simple ensemble of two-level or multilevel systems with equal transition energies [5]. In a single-particle picture, the eigenstates of a semiconductor are characterized by a full valence band and an empty conduction band, both with a continuous density of states. The photoexcitation of such a system with a broad laser pulse can be described by an inhomogeneous superposition of two-level systems, which leads to a rapid dephasing due to polarization interference. The polarization can be recovered in a photon echo experiment when a suitable sequence of laser pulses leads to a reversal of the polarization vector motion which rephases the polarization. This photon echo picture of the dephasing of continuum states in a semiconductor, which was used to explain the results of first studies of the phase relaxation dynamics of a semiconductor at room temperature [10], has recently been confirmed by time-resolved four-wave-mixing experiments, where the photon echo was experimentally observed [11].

The corresponding theoretical description of the interband transitions in a semiconductor employs a density matrix which characterizes a continuum of electron and hole states described by the k-vector(s) of the excitations. The equations for the polarization and population of an individual transition read [12]

$$\frac{\partial}{\partial t}p_k = \frac{\partial}{\partial t}p_k\big|_{\text{coh}} + \frac{\partial}{\partial t}p_k\big|_{\text{scatt}} \tag{3.16}$$

for the terms relevant to the polarization, and

$$\frac{\partial}{\partial t}n(k)_{\text{e,h}} = \frac{\partial}{\partial t}n(k)_{\text{e,h}}\big|_{\text{coh}} + \frac{\partial}{\partial t}n(k)_{\text{e,h}}\big|_{\text{scatt}} \tag{3.17}$$

for the occupation numbers $n_{\text{e}}(k)$ and $n_{\text{h}}(k)$. The optical Bloch equations, with the multitude of k states taken into account in the form outlined above, are the semiconductor Bloch equations in their simplest form.

A detailed discussion of the properties of optically excited semiconductors should also take the Coulomb many-body interaction between the carriers into account. In the simplest picture, the excitation spectrum of the semiconductor is characterized by excitons and their continuum states: the lowest excited state of a semiconductor is the exciton $n = 1$ state. Above that, excited exciton states and the exciton continuum are present. The exciton continuum can then be linked to the free electrons and holes in the single-particle picture of a semiconductor.

From an experimentalist's point of view, the existence of the Coulomb interactions has two main effects with regard to the optical excitation of Bloch oscillations:

- A detrimental effect is that the Coulomb interaction leads to many pronounced changes in the optical response. In particular, for the multilevel systems which are the main topic of this book, Coulomb coupling modifies the simple single-particle textbook-like structures. Also, the nonlinear optical properties of the systems are modified.
- A favorable effect is that the Coulomb interaction leads to excitons, which in many cases can be described quite well by the simple two-level-system model. It is thus possible to describe a large number of nonlinear optical experiments in semiconductors using the optical Bloch equations formulated for two-level systems.

It is beyond the scope of this section to review the various effects of Coulomb interaction on the linear and nonlinear optical processes in semiconductors (for a lucid discussion in an experimentalist's language, consult [12]). We briefly mention here that theoretical models have been derived which describe the experimental results quite well. For the linear optical properties, variational solutions of the Schrödinger equation including the Coulomb coupling usually describe the properties of many-level systems in an appropriate manner (see, e.g., [13] for the case of the resonant coupling of quantum wells). For the nonlinear properties, it turns out that it is necessary to include both the external field and the Coulomb-induced coupling in the driving terms of the polarization [12,14,15]. Although the effects due to polarization coupling are visible in some of the data discussed here, we can in most cases avoid a more detailed discussion since those effects do not influence the interpretation of the Bloch oscillation experiments.

However, we shall briefly discuss here results of Leisching et al. [16], which have directly demonstrated that the Coulomb-induced coupling is the origin of the polarization-induced scattering. These authors used the fact that the Coulomb coupling in the Wannier–Stark ladder becomes progressively weaker with increasing field owing to the larger k-mismatch. Therefore, the temporal dependence of the polarization in a biased superlattice can be used to investigate the polarization dynamics as a function of the Coulomb coupling. To detect this temporal dependence Leisching et al. [16] measured the time-resolved third-order polarization in a superlattice when Bloch oscillations were excited. The data clearly demonstrated that for low fields, i.e. strong coupling, the polarization showed a pronounced delayed rise due to the buildup of the polarization wave, whereas for higher fields, the buildup was much faster, indicating that the influence of polarization coupling was then weaker.

It has recently been shown [17] that the semiconductor Bloch equations are inadequate to describe experiments where inter- and intraband polarizations play a role. The path beyond the semiconductor Bloch equations involves the inclusion of all important higher-order correlation terms. One area where this approach is relevant to the discussion here is the comparison of interband and intraband dynamics. Theoretical approaches based on the dynamically controlled truncation (DCT) model [17] predict that the intraband dynamics are influenced by excitonic effects in a similar way to the interband dynamics. A recent comparison of THz spectroscopy and four-wave-mixing results has confirmed this prediction [18].

3.3 Optical Techniques to Detect Bloch Oscillations

We follow here a recent overview of those optical techniques which are capable of tracing coherent excitations [5]. In most experiments, coherent ensembles of carriers have been generated using optical techniques. Illumination with a coherent laser source leads to the phase-correlated generation of a superposition of electron and hole states or, in more theoretical language, to the generation of phase-related electron–hole pair amplitudes. For the detection of the coherently generated excitations, both *interband* and *intraband* optical techniques have been used. In the interband case, excitations above the band gap are probed to investigate the dephasing and/or the interference dynamics of previously excited states. In the intraband case, the radiation of the sample due to interference phenomena is probed at the far-infrared frequencies which are typical of Bloch oscillations.

It should be noted that it has recently become possible to use optically excited THz radiation to pump coherent transitions in the intraband range and monitor the polarization decay (see, e.g., [19]). With the availability of high-power sources such as free-electron lasers (FELs), it can be expected that the full spectrum of nonlinear methods used for interband experiments will also become available in the intraband range. For instance, four-wave mixing at intraband frequencies will give valuable information about the intraband dephasing times, avoiding the influence of inhomogeneous broadening and the photoexcitation of holes.

In the following, we give a brief overview of the optical techniques which can be used to study coherent transport in superlattices.

3.3.1 Interband Optical Techniques

Owing to the generally short lifetime of coherent excitations in semiconductors, optical techniques need to be able to resolve signals in the subpicosecond regime. With few exceptions, the techniques used have therefore been correlation techniques where a first ultrafast optical pulse excites the sample and sets up the coherent polarization, and a second pulse impinges after a well-defined delay time to probe the time evolution of this polarization.

The two correlation techniques which have been frequently used are four-wave mixing and transient absorption (pump–probe). In both techniques, two laser beams, with directions $\boldsymbol{k}_1$ and $\boldsymbol{k}_2$, are focused onto the sample at an angle, as shown in Fig. 3.2. The first laser pulse generates a coherent polarization, which radiates in the direction $\boldsymbol{k}_1$.

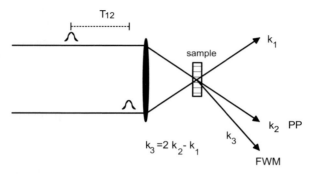

Fig. 3.2. Schematic illustration of a two-beam experiment with detection at $2\boldsymbol{k}_2 - \boldsymbol{k}_1$ (four-wave mixing) and $\boldsymbol{k}_2$ (pump–probe)

For the pump–probe experiment, the detection is in the direction of the probe beam $\boldsymbol{k}_2$. It can be shown that the change of this signal due to the presence of the pump pulse $\boldsymbol{k}_1$ can be divided into two parts. First, there are signal contributions which are of incoherent nature and can be classified as phase space filling and screening. Additionally, there is a coherent signal which is due to the third-order polarization. The fact that the pump–probe experiment has contributions from both incoherent and coherent excitations is a disadvantage if one wants to concentrate on the nature of coherent states. A second disadvantage of the pump–probe experiment is that the coherent part of the signal is influenced by a polarization decay due to inhomogeneous broadening since there is no rephasing. It is thus not possible to investigate the coherent dynamics of systems with strong inhomogeneous broadening.

Most experiments which have addressed coherent dynamics in semiconductors have used four-wave mixing. The popularity of this technique for detecting coherent excitations in semiconductors results basically from the facts that (i) it detects only coherent contributions to the optical signal, thus suppressing the strong incoherent response of semiconductors and (ii) since the signals are detected in a background-free direction, stray light due to surface imperfections of the sample is suppressed. A detailed theoretical study of four-wave mixing in biased superlattices was performed by von Plessen and

Thomas [20]. Previously, Zakharov and Manykin had treated the four-wave-mixing response of a bulk semiconductor subject to an electric field [21].

Figure 3.2 also shows the detection geometry of a typical four-wave-mixing and pump–probe setup. This diagram describes a particularly simple version of four-wave mixing, using the so-called self-diffracted geometry [22], usually performed with two laser pulses with identical spectra, i.e. degenerate frequencies. The temporal delay T_{12} between the two pulses can be varied by sending either one of them over a variable delay line. As in the case of the pump–probe experiment, the pulse which arrives first sets up a macroscopic polarization $P^{(1)}(t)$ in the sample. The second pulse then acts twice. First, its electric field interferes with $P^{(1)}(T_{12})$ and creates a real population of carriers and, second, it generates a third-order polarization $P^{(3)}(\omega, t, T_{12})$. This nonlinear polarization is the source of the four-wave-mixing signal emitted in the direction

$$\boldsymbol{k}_3 = 2\boldsymbol{k}_2 - \boldsymbol{k}_1 \ . \tag{3.18}$$

This signal can arise from a variety of nonlinear optical effects such as phase space filling, screening, excitation-induced dephasing, and band gap renormalization. In simple terms, photons of the second pulse are diffracted from a grating set up by the first and the second pulse, with a grating vector $\boldsymbol{k}_2 - \boldsymbol{k}_1$.

In the four-wave-mixing experiment, the third-order polarization emitted in the direction $\boldsymbol{k}_3$ can be detected in different ways (see Fig. 3.3):

- time-integrated detection using a slow detector without spectral resolution;
- spectrally resolved detection by means of a slow detector coupled to a spectrometer;

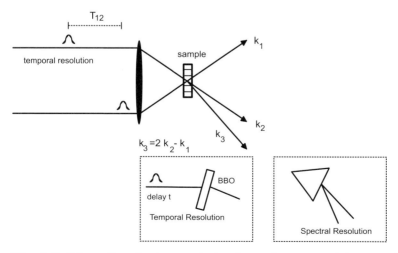

Fig. 3.3. Schematics of a degenerate two-beam four-wave mixing experiment with the various detection schemes.

- time-resolved detection, mostly done via sum frequency generation by mixing the signal in a nonlinear crystal with a third, strong, reference pulse.

The time t in the time-resolved scheme is the real time at which the third-order polarization is detected. For a more extensive description of the technique, we refer to [22–26].

A theoretical analysis of transient four-wave-mixing signals has been performed by Yajima and Taira [22] by solving the optical Bloch equations to third order for an ensemble of noninteracting two-level systems. These authors found that the decay of the four-wave-mixing signal is related to the interband dephasing time T_2^{inter}. In the limit of homogeneous broadening of the transition, the intensity decays according to

$$I(T_{12}) \propto \exp\left(\frac{-2T_{12}}{T_2^{\text{inter}}}\right). \tag{3.19}$$

If the transition is inhomogeneously broadened and is formed by a superposition of many transitions, typically described by a Gaussian distribution of transition energies with a width much larger than the homogeneous broadening, the decay is given by

$$I(T_{12}) \propto \exp\left(\frac{-4T_{12}}{T_2^{\text{inter}}}\right). \tag{3.20}$$

For cases where homogeneous and inhomogeneous broadening are of comparable importance, the Bloch equations have to be solved numerically to obtain the relation between the four-wave-mixing decay time and the dephasing time. Note that a two-beam four-wave-mixing experiment does not allow one to distinguish between homogeneous and inhomogeneous broadening. To do this, one has two possibilities:

- In a three-beam four-wave-mixing experiment, the diffracted signal, as a function of the delay between the first two pulses, is symmetric for homogeneous broadening and asymmetric for inhomogeneous broadening [27].
- In a two-beam experiment with time-resolved detection of the diffracted beam, one should observe a photon echo for inhomogeneous broadening and an exponential decay for homogeneous broadening. However, the temporal behavior is influenced by polarization scattering, so that this conclusion cannot always be drawn [28].

For the observation of quantum interference phenomena, both of the interband techniques pump–probe and four-wave mixing can be used. In a pump–probe experiment, the transmitted signal oscillates as a function of the delay time. However, since there are usually incoherent contributions to the pump–probe signal, the coherent response might be only a small part of the observed signal intensity.

Similarly, interference can be observed in time-integrated four-wave-mixing by detecting the diffracted signal as a function of T_{12}. Quantum beats are also observed in the time-resolved emission, i.e. as a function of

t for a fixed time delay T_{12}. The spectrally resolved four-wave-mixing signal for each time delay T_{12} shows distinct resonances which relate to the photoexcited transitions.

To describe the response in an experiment where Bloch oscillations are being investigated, we again discuss here the response for the simultaneous excitation of the two upper states of a three-level system [29]. Again, the ground state is the level $|g\rangle$; the two excited states are denoted by $|e1\rangle$ and $|e2\rangle$. The damping rates of the interband coherence are $\gamma_{g,e1}$ and $\gamma_{g,e2}$, and the intraband relaxation rate is $\gamma_{e1,e2}$. The signals for the pump–probe and four-wave-mixing experiments are calculated using a density matrix theory with a perturbation ansatz. The pump–probe and four-wave-mixing signals $I_{\rm PP}$ and $I_{\rm FWM}$ are then proportional to the third-order polarization in the directions $\boldsymbol{k}_2$ and $2\boldsymbol{k}_2 - \boldsymbol{k}_1$, respectively. For δ-shaped pulses and purely homogeneous broadening, one obtains the following relations [29]:

$$I_{\rm PP} \propto w_1^2 + w_2^2 + w_1 w_2 \left[1 + \cos(\Delta E_{12} t)\exp(-\gamma_{e1,e2} t)\right] \qquad (3.21)$$

and

$$\begin{aligned} I_{\rm FWM} \propto\; & w_1^2 e^{-2\gamma_{g,e1} t} + w_2^2 e^{-2\gamma_{g,e2} t} \\ & + 2 w_1 w_2 \cos(\Delta E_{12} t)\exp\left[-(\gamma_{g,e1} + \gamma_{g,e2}) t\right] \;, \end{aligned} \qquad (3.22)$$

where w_1 and w_2 are the spectral weights of the transitions and ΔE_{12} is the energetic splitting between the excited states. The most important conclusion is that the *intraband phase relaxation* $\gamma_{e1,e2}$ between the excited-state levels is responsible for the damping of the oscillations in the pump–probe experiment, i.e. the pump–probe experiment probes the intraband dephasing, which is responsible for the Bloch oscillation decay. In contrast, the quantum beat oscillations in the four-wave-mixing experiment probes the sum of the *interband phase relaxation rates* to the ground states, $\gamma_{g,e1} + \gamma_{g,e2}$.

It is interesting to note that a numerical evaluation of (3.22) shows that the overall exponential decay of the four-wave-mixing signal is modulated with an oscillation of constant modulation depth if the two relaxation rates $\gamma_{g,e1}$ and $\gamma_{g,e2}$ are equal [29], i.e. the oscillation persists undamped (relative to the exponentially decaying overall signal). Also, if these two rates are different, the damping of the modulation of this oscillation is proportional to the difference between the two rates $\gamma_{g,e1}$ and $\gamma_{g,e2}$. If one assumes that the difference is caused by some scattering process between levels $|e1\rangle$ and $|e2\rangle$ (and not by coupling to a separate bath), then the four-wave-mixing experiment also reveals information about the intraband damping.

It should be mentioned here that the two-beam four-wave-mixing technique delivers considerably less information than does the three-beam technique (see Fig. 3.4), where all three incoming light beams have different directions. As pointed out in [27], the three-beam technique allows one to independently measure the polarization and population decays. In this experiment, three beams with directions $\boldsymbol{k}_1$, $\boldsymbol{k}_2$, and $\boldsymbol{k}_3$ impinge on the sample. The nonlinear signal is measured in the direction

$$k_4 = k_3 + (k_2 - k_1) \ . \tag{3.23}$$

The experiment can be understood in simple terms as a transient-grating experiment. The first pulse sets up a polarization of first order, and the second pulse forms a grating with the polarization left over from the first pulse. The third pulse is then diffracted from this grating (see Fig. 3.5 for

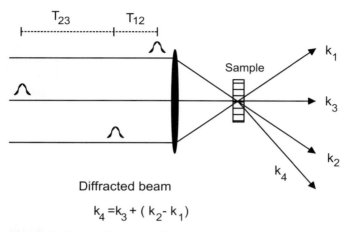

Fig. 3.4. Schematic illustration of a three-beam four-wave-mixing experiment

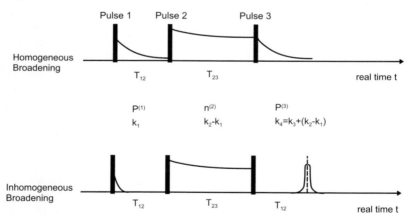

Fig. 3.5. Dynamics of polarization and population in a three-beam four-wave-mixing experiment for homogeneous (*top*) and inhomogeneous (*bottom*) broadening. The first pulse generates in both cases a polarization $P^{(1)}$ in direction k_1, which decays much faster for the case of inhomogeneous broadening. The second pulse then leads to a slowly decaying population, which is probed by the third pulse, resulting in a third-order polarization $P^{(3)}$ which slowly decays for homogeneous broadening or is emitted as a photon echo for inhomogeneous broadening

a schematic description). By variation of the delay time T_{12} between the first two pulses with T_{23} kept constant, the decay of the polarization of the first pulse can be determined. By variation of the delay time T_{23} between the second and the third pulse (with T_{12} kept constant), the decay of the grating due to processes such as recombination or diffusion can be monitored.

As mentioned above, the three-beam experiment also allows one to distinguish between homogeneous and inhomogeneous broadening: the signal, as a function of the delay T_{12} between the first two pulses, is symmetric around zero delay for homogeneously broadened transitions and asymmetric for inhomogeneously broadened transitions (see [30] for an application). In simple terms, this asymmetry is caused by the fact that the polarization needs to be rephased in the case of inhomogeneous broadening, which works only if the time sequence is correct since the rephasing is otherwise in the wrong direction. For homogeneous dephasing, the time sequence does not matter.

3.3.2 Other Techniques

In the last few years, time-resolved spectroscopic techniques in the spectral range of the intraband transitions have been developed (for a review, see the book chapter by Nuss and Orenstein [31]). The simplest realization of an experiment which detects Bloch oscillations is one in which a first laser pulse generates an oscillating wave packet. This wave packet will then radiate in the THz range as long as the intraband coherence is preserved. Using, for example, a photoconductive antenna, with which the THz field is detected as a function of the delay time of a second optical gating pulse, one can directly trace the dynamics of the THz field. This technique will be described in more detail in Chap. 5 together with the respective experiments.

Another technique which has been used to detect Bloch oscillations is transmittive electro-optic sampling (TEOS) (see Fig. 3.6). This technique uses the birefringence induced by an electric field: if an electric field is applied in the z direction [001] of the superlattice, the index of refraction seen by a light beam propagating in the [001] direction becomes dependent on its polarization. Owing to the concomitant phase shift during propagation through the sample, a circularly polarized probe beam will become elliptically polarized. This change in polarization can be detected by a polarized beam splitter and two detectors. If the electro-optic coefficient is known, it is possible to measure internal fields absolutely. This effect can therefore be used to measure the internal oscillating field caused by the Bloch oscillations and thus obtain the amplitude of the Bloch wave packet when the carrier density is known. The TEOS technique has been used with resonant (near-band-gap) [32] and off-resonant [33] excitation to study Bloch oscillations. As discussed in detail in Sect. 4.1, there are problems with resonant excitation, since a quantitative analysis is hampered and other signal contributions might be much larger.

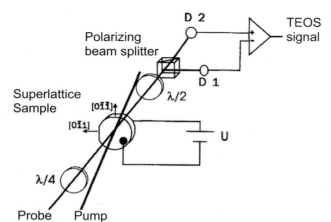

Fig. 3.6. Schematic illustration of a transmittive electro-optic sampling experiment

3.4 Observation of Bloch Oscillations by Transient Four-Wave Mixing

As discussed above, the most straightforward way to observe the coherent ensemble motion associated with Bloch oscillations is to excite the system with a short laser pulse, the width of which is larger than the spacing between two adjacent levels of the Wannier–Stark ladder, to set up a wave packet of Wannier–Stark ladder states. This is equivalent to the condition that the optical pulse is shorter than the period of a Bloch oscillation. It should be pointed out that the simple observation of these quantum beats by four-wave mixing is a corollary of the observation of the Wannier–Stark ladder in frequency space. However, one difference is that Bloch oscillations can be observed in transient four-wave-mixing experiments even if the Wannier–Stark ladder is not observed: for strong inhomogeneous broadening of the system, Bloch oscillations might be observed despite the fact that the Wannier–Stark ladder was smeared out in the optical spectra. In other words, the four-wave-mixing experiment yields information about homogeneous broadening mechanisms which cannot be obtained from linear spectra, since the linear spectra are influenced by both homogeneous and inhomogeneous broadening.

We were stimulated to do the first experiments on Bloch oscillations using four-wave mixing by a talk by Gerald Bastard, who had performed a calculation of the spatial dynamics of Wannier–Stark states with the special assumption that the initial state is localized in one well [35]. One can show that this initial condition corresponds to a wave packet which is spread in k-space and thus performs a breathing-mode oscillation without center-of-mass motion. After the discovery [29] of coherent oscillations in coupled quantum wells (which was, in a way, the most basic spatial-wave-packet experiment in a heterostructure), we set out to extend those experiments to a structure with many wells. The first design used for this purpose was a

3.4 Observation of Bloch Oscillations by Transient Four-Wave Mixing

GaAs/Al$_{0.3}$Ga$_{0.7}$As superlattice structure with 20 periods of 3 nm barriers and wells. In the center, an enlarged well of width 4.2 nm was placed, in order to allow selective excitation in one well only, in accordance with the proposal by Bastard and Ferreira [35]. A tandem-pumped dye laser with a pulse duration of about 500 fs was used as laser source. However, it turned out that this structure did not give a sufficient four-wave-mixing signal to observe the effect. Unfortunately, the idea of using an enlarged well to start a Bloch oscillation wave packet has not been pursued further, although it would be a quite nice demonstration experiment.

The first experiments which indicated the existence of Bloch oscillations were performed some time later on a standard superlattice structure [34]. The standard two-beam transient four-wave-mixing technique [22] was employed as experimental technique. The sample was a 97 Å/17 Å Al$_{0.3}$Ga$_{0.7}$As superlattice embedded in a pin-structure and processed as a mesa structure. During the experiments, the sample was held in a cryostat at approximately 10 K.

The Wannier–Stark ladder and the four-wave mixing data are depicted in Fig. 3.7. In the experiments, the bias voltages had to be carefully adjusted to

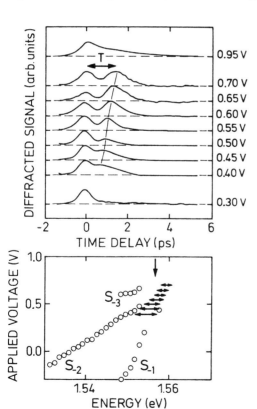

Fig. 3.7. Lower part: Wannier–Stark ladder for a 97 Å/17 Å superlattice. The optical excitation is indicated by *arrows*; Upper part: Four-wave-mixing signal in the same bias voltage region (from [34])

compensate for the influence of the screening of the photogenerated carriers on the effective electric field in the superlattice region. The optical excitation was chosen in a way that the laser was exciting both the hh_{-1} and the hh_{-2} transition; owing to the rather small band width of the tandem-pumped dye laser system [36] of about 4 meV, the center frequency of the laser had to be adjusted with increasing field (arrows in Fig. 3.7, lower part) to excite a wave packet of the two states of the Wannier–Stark ladder.

The experimental data showed hints of Bloch oscillations, but were not fully convincing: they showed a recovery of the four-wave-mixing signal after a time period which was in agreement with the tunable inverse Wannier–Stark ladder splitting. However, for reasons not entirely understood, no further oscillations were visible (see also Fig. 3.8).

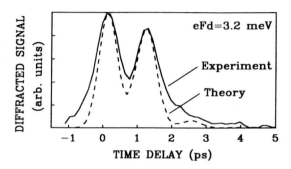

Fig. 3.8. Four-wave mixing signal for a biased 97 Å/17 Å superlattice (from [34])

The first clear observation of Bloch oscillation signals with several cycles was obtained for a similar superlattice structure by four-wave-mixing experiments [37]. Figure 3.9 shows the Wannier–Stark ladder observed by photocurrent spectroscopy for that sample. Several transitions of both the heavy-hole and the light-hole ladder are clearly visible.

The time-resolved data were taken at approximately 10 K with a 100 fs Ti-sapphire laser. In the low-field range, the four-wave-mixing response showed quantum beats of the heavy-hole and light-hole excitons. For higher fields, oscillations with a period inversely proportional to the bias voltage (i.e. to the electric field) were visible. The four-wave-mixing scans for given fields showed oscillations with a period which was tunable over a large field range [37]: Figure 3.10 shows scans of the four-wave-mixing signal vs. delay for various electric fields. It is obvious that the electronic wave packets can perform several Bloch oscillations before they are damped. The damping time is obviously only weakly, if at all, dependent on the electric field, indicating that transitions to higher bands (Zener tunneling) do not play a strong role in the damping.

It is interesting to note that the four-wave-mixing signal extends to negative delay times and also shows oscillatory parts there. This negative signal, which is not expected from the simple explanation of four-wave mixing dis-

3.4 Observation of Bloch Oscillations by Transient Four-Wave Mixing   45

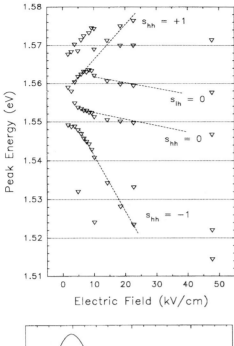

Fig. 3.9. Optical transition energies taken from photocurent spectra of a 97Å/17Å superlattice taken at 10 K, showing the fanchart associated with the Wannier–Stark ladder (from [37])

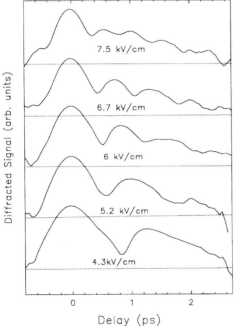

Fig. 3.10. Four-wave mixing signal of a 97Å/17Å superlattice for various fields (from [37])

cussed above, is actually well understood and is caused by local-field effects: in a many-particle system, the grating can scatter not only the photons of the second pulse (or the third pulse, in the case of three-beam four-wave mixing) in a background-free direction, but also the polarization left over from the first pulse. It is easy to understand that this scattering can occur only if the system is dominated by homogeneous broadening; otherwise, the first-order polarization of the first pulse would disappear very quickly. It was shown that this negative-delay-time signal decays twice as fast as the positive-delay signal, in agreement with theory [14].

Figure 3.11 shows the splitting E_{BO}, calculated from the experimentally measured period using (1.10), plotted vs. the electric field. The agreement of the oscillation period with theory is good. Note that the tuning range extends from 2 to 10 meV in terms of the splitting, i.e. an oscillation frequency range of about 0.5 to 2.5 THz. It should be noted here that this large tuning range is very remarkable for a quantum oscillator. The tuning range of a Bloch oscillator is, to the author's knowledge, by far the largest tuning range of an elementary oscillator observed in any system.

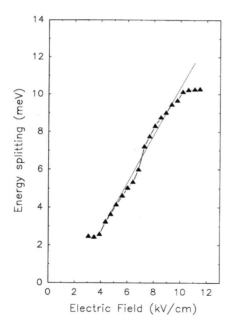

Fig. 3.11. Energy splitting derived from the period of Bloch oscillations in a 97 Å/17 Å superlattice vs. electric field (from [37])

The time-resolved experiments discussed here are strongly influenced by the screening of the static bias field due to the photogenerated carriers: the transitions of the Wannier–Stark ladder are shifted in bias voltage compared with continuous-wave measurements since the photogenerated carriers created by the pulsed excitation create a dipole moment which reduces the static field in the pin junction. A comparative measurement of the Wannier–Stark ladder under continuous-wave and pulsed optical excitation [39] has shown that:

- The screening is a slow phenomenon compared with the repetition rate of the pulsed laser system (80 MHz). The Wannier–Stark ladder, under pulsed excitation, displays sharper transitions than under continuous-wave excitation, but the transitions are shifted in terms of the applied voltage. The additional bias due to the photogenerated carriers is thus quasi-static. However, the calibration of the field has to be done for each experiment separately since the dynamics is probably determined by the superlattice and the electrical circuit connected to it.
- The screening due to the photogenerated carriers causes a shift of the Wannier–Stark ladder in terms of the bias voltage; once a certain threshold is reached, the transitions scale with the field as in the continuous-wave case.

In the data presented in Figs. 3.10 and 3.11, the screening of the electric field in terms of pulsed laser excitation has accordingly been taken into account by subtracting a constant voltage offset.

3.5 Further Investigations with Four-Wave Mixing

After the initial experiments leading to the observation of Bloch oscillations, a number of further studies using four-wave mixing as the experimental technique have helped to improve the understanding of Bloch oscillations. Among those studies, a simple but important experiment provided a clear-cut demonstration that the observed effects were indeed due to Bloch oscillations: it is an important point to show that the oscillatory signal observed in the four-wave-mixing experiments is due to quantum interference and not caused by polarization interference effects, i.e. oscillations of the signal due to a far-field interference of the signals radiated from the various Wannier–Stark ladder transitions.

3.5.1 Quantum vs. Polarization Oscillations

In the standard transient four-wave-mixing experiments, one cannot distinguish conclusively whether the oscillation observed is a true quantum mechanical interference within a wave packet (as is the case for Bloch oscillations) or whether it is caused by a polarization interference. Even in THz emission experiments (discussed in Chap. 5), one cannot exclude the possibility that the emission is due to the nonlinear mixing of two such polarizations.

The distinction between quantum and polarization interference is most easily and conclusively performed with spectrally resolved four-wave mixing experiments. Tokizaki et al. [38] have shown that spectral resolution of the diffracted beam can directly distinguish between two oscillation mechanisms: for quantum oscillations, the phase of the oscillations is independent of the detection energy, but for polarization oscillations, the phase depends on the detection energy.

Experiments by Leisching et al. [39] have shown unambiguously that the phase of the oscillations observed in four-wave mixing is not dependent on the detection energy, thus proving that the oscillations are quantum oscillations, as expected for Bloch oscillations. Figure 3.12 shows spectrally resolved four-wave-mixing data for a 67 Å/17 Å $Al_{0.3}Ga_{0.7}As$ sample. The phase of the oscillations is independent of the detection energy, although the absolute four-wave-mixing signal varies strongly as a function of energy owing to the strong spectral variation of the nonlinearity that shows the Wannier–Stark ladder transitions. These data thus prove that the oscillations are caused by a quantum interference, as should be the case if Bloch oscillations are observed.

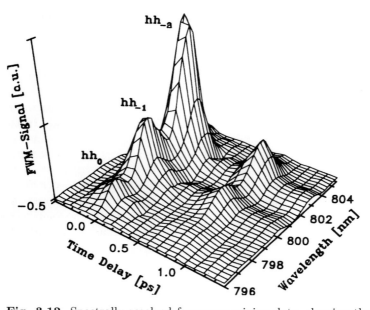

Fig. 3.12. Spectrally resolved four-wave-mixing data, showing that the phase of the oscillatory signal is independent of the detection energy (from [39])

3.5.2 Influence of Miniband Width

One of the key advantages of semiconductor superlattices is the ability to vary the miniband width over a large range. It is important to investigate whether and how the dynamic phenomena depend on the width of the band. Also, the tunability of the Bloch oscillations is related to the ratio Δ/eFd of the miniband width to the field energy, i.e., for large tunability, broad minibands are needed.

Using optical experiments, such studies can be performed over a wide energy range. The lower limit is given by a miniband width of a few meV, which starts to fall in the region of the exciton binding energy. In this regime, there are still quantum oscillations observable, as shown experimentally [40]. However, the linear tunability with bias field is lost, since the excitonic coupling dominates over the effects due to the Wannier–Stark ladder formation [13]. The upper limit for the investigation of the miniband width is given by the laser bandwidth or by the band offset of the chosen heterostructure system.

It was claimed in Ref. [41], from theoretical arguments, that a miniband width equal to the optical-phonon energy (which is on the order of 36 meV in GaAs superlattices) would be a threshold at which strong damping would set in. These authors also presented data for such a system which did not show any oscillatory signature and claimed that this was caused by the strong damping by LO phonons when the miniband width exceeded this threshold.

This conclusion, however, has been invalidated by more recent experiments. Bloch oscillations in samples with a miniband width larger than the LO phonon energy have been observed [39], even at splittings eFd smaller than that reported in the earlier study [41]. Leisching et al. [39] have shown by four-wave-mixing experiments, that Bloch oscillations exist in $GaAs/Al_{0.3}Ga_{0.7}As$ samples with miniband widths ranging from 19 to 50 meV. A pronounced influence of the miniband width on the Bloch oscillations was not observed, although the dephasing times decreased with increasing miniband width.

The question of the influence of the miniband width on Bloch oscillations is closely related to the more general problem of the damping of Bloch oscillations by scattering, which will be discussed in detail in Sect. 6.2. The influence of the miniband width on the electrical transport properties of superlattices will be discussed in Chap. 8.

3.6 Summary

In this chapter, we have discussed the optical studies with interband excitation and detection which led to the discovery of Bloch oscillations. The key observation is the occurrence of a beating of Wannier–Stark ladder states with a frequency which is widely tunable through the electric field. Further experiments have shown that this beating is indeed a quantum oscillation of an optically generated wave packet and that the oscillations exist over a large range of miniband width, even well above the optical-phonon energy.

References

1. K. Leo, Semicond. Sci. Technol. **13**, 249 (1998).
2. V.G. Lyssenko, M. Sudzius, F. Löser, G. Valusis, T. Hasche, K. Leo, M.M. Dignam, K. Köhler, Festkörperprobleme/Adv. Solid State Phys. **38**, 225 (1999).
3. K. Leo and M. Koch, in *The Physics and Chemistry of Wave Packets*, ed. J. Yeazell and T. Uzer (Wiley, New York 2000), p. 285.
4. P. Haring Bolivar, T. Dekorsy, and H. Kurz, in *Semiconductors and Semimetals*, Vol. 66, ed. G. Mueller (Academic Press, San Diego 2000), p. 187.
5. K. Leo, "Coherent Effects in Semiconductor Heterostructures", in *Handbook of Advanced Electronic and Photonic Materials and Devices*, ed. H.S. Nalwa (Academic Press, San Diego 2001), p. 279.
6. D. Dunlap, private communication.
7. F. Rossi, Semicond. Sci. Technol. **13**, 147 (1998).
8. M.M. Dignam, J.E. Sipe, and J. Shah, Phys. Rev. B **49**, 10502 (1994).
9. A.M. Bouchard and M. Luban, Phys. Rev. B **47**, 6815 (1993).
10. P.C. Becker, H.L. Fragnito, C.H. Brito Cruz, R.L. Fork, J.E. Cunningham, J.E. Henry, and C.V. Shank, Phys. Rev. Lett. **61**, 1647 (1988).
11. W.A. Hügel, M.F. Heinrich, M. Wegener, Q.T. Vu, L. Banyai, and H. Haug, Phys. Rev. Lett. **83**, 3313 (1999).
12. D.S. Chemla, in *Semiconductors and Semimetals*, Vol. 58, eds. E.R. Weber and K.K. Willardson (Academic Press, San Diego 1998), p. 175.
13. A.M. Fox, D.A.B. Miller, G. Livescu, J.E. Cunningham, J.E. Henry, and W.Y. Yan, Phys. Rev. B **42**, 1841 (1990).
14. K. Leo, M. Wegener, J. Shah, D.S. Chemla, E.O. Göbel, T.C. Damen, S. Schmitt-Rink, and W. Schäfer, Phys. Rev. Lett. **65**, 1340 (1990).
15. M. Wegener, D.S. Chemla, S. Schmitt-Rink, and W. Schäfer, Phys. Rev. A **42**, 5675 (1990).
16. P. Leisching, W. Beck, H. Kurz, W. Schäfer, K. Leo, and K. Köhler, Phys. Rev. B **51**, R7962 (1995).
17. V.M. Axt, G. Bartels, and A. Stahl, Phys. Rev. Lett. **76**, 2543 (1996).
18. P. Haring Bolivar, F. Wolter, A. Müller, H.G. Roskos, H. Kurz, and K. Köhler, Phys. Rev. Lett. **78**, 2232 (1997).
19. J.N. Heyman, R. Kersting, and K. Unterrainer, Appl. Phys. Lett. **72**, 644 (1998).
20. G. von Plessen and P. Thomas, Phys. Rev. B **45**, 9185 (1992).
21. S.M. Zakharov and E.A. Manykin, Izv. Akad. Nauk. SSSR, Ser. Fiz. **37**, 2171 (1973).
22. T. Yajima and Y. Taira, J. Phys. Soc. Jpn. **47**, 1620 (1979).
23. J. Shah, *Ultrafast Spectroscopy of Semiconductors and Semiconductor Nanostructures*, 2nd ed. (Springer, Berlin, Heidelberg 1999).
24. J. Kuhl, A. Honold, L. Schultheis, and C.W. Tu, Festkörperprobleme/Adv. Solid State Phys. **29**, 157 (1989).
25. Y.R. Shen, *The Principles of Nonlinear Optics* (Wiley, New York 1984).
26. M. Koch, G. von Plessen, J. Feldmann, and E.O. Göbel, Chem. Phys. **120**, 367 (1996).
27. A.M. Weiner, S. De Silvresti, and E.P. Ippen, J. Opt. Soc. Am. B **2**, 654 (1985).
28. D.S. Kim, J. Shah, T.C. Damen, W. Schäfer, F. Jahnke, S. Schmitt-Rink, and K. Köhler, Phys. Rev. Lett. **69**, 2725 (1992).

29. K. Leo, J. Shah, E.O. Göbel, T.C. Damen, S. Schmitt-Rink, W. Schäfer, and K. Köhler, Phys. Rev. Lett. **66**, 201 (1991).
30. K. Leo, Appl. Phys. A **53**, 118 (1991).
31. M.C. Nuss and J. Orenstein, in *Millimeter and Submillimeter Wave Spectroscopy of Solids*, ed. G. Grüner (Springer, Berlin, Heidelberg 1998), p. 7.
32. T. Dekorsy, P. Leisching, K. Köhler, and H. Kurz, Phys. Rev. B **50**, 8106 (1994).
33. M. Först, G. Segschneider, T. Dekorsy, H. Kurz, and K. Köhler, Phys. Rev. B **61**, R10563 (2000).
34. J. Feldmann, K. Leo, J. Shah, D.A.B. Miller, J.E. Cunningham, S. Schmitt-Rink, T. Meier, G. von Plessen, A. Schulze, and P. Thomas, Phys. Rev. B **46**, 7252 (1992).
35. G. Bastard and R. Ferreira, in *Spectroscopy of Semiconductor Microstructures*, Vol. 206 of NATO Advanced Study Institute, Series B: Physics, Eds. G. Fasol and A. Fasolino (Plenum, New York 1989), p. 333.
36. K. Leo, J. Shah, T.C. Damen, A. Schulze, T. Meier, S. Schmitt-Rink, P. Thomas, E.O. Göbel, S.L Chuang, M.S.C. Luo, W. Schäfer, K. Köhler, P. Ganser, IEEE J. Quantum Electronics QE **28**, 2498 (1992).
37. K. Leo, P. Haring Bolivar, F. Brüggemann, R. Schwedler, and K. Köhler, Solid State Commun. **84**, 943 (1992).
38. T. Tokizaki, A. Nakamura, Y. Ishida, T. Yahima, I. Akai, and T. Karasawa, in *Ultrafast Phenomena VII*, eds. C.B. Harris, E.P. Ippen, G.A. Mourou, and A.H. Zewail, Springer Series in Chemical Physics, Vol. 53 (Springer, Berlin 1990), p. 253.
39. P. Leisching, P. Haring Bolivar, W. Beck, Y. Dhaibi, F. Brüggemann, R. Schwedler, H. Kurz, K. Leo, and K. Köhler, Phys. Rev. B **50**, 14389 (1994).
40. G. Cohen, I. Bar-Joseph, and H. Shtrikman, Phys. Rev. B **50**, 17316 (1994).
41. G. von Plessen, T. Meier, J. Feldmann, E.O. Göbel, P. Thomas, K.W. Goossen, J.M Kuo, and R.F. Kopf, Phys. Rev. B **49**, 14058 (1994).

4 Interband Optical Experiments on Bloch Oscillations: Measurements of the Spatial Dynamics

In the previous chapter, we discussed the first interband optical experiments on Bloch oscillations, which clarified the basic phenomena. Although those experiments have brought important insight, they have only offered information about the temporal dynamics of the Bloch oscillations, which is, in principle, only a corollary of the observation of the Wannier–Stark ladder in frequency space.

In this chapter, we discuss more recent experiments which go further. The emphasis is laid on experiments which use a novel experimental technique that allows one to directly measure the spatial displacement of the carriers as a function of time, and thus to investigate the properties of the harmonic motion predicted by Zener (Sect. 4.1). We then discuss experiments which show that the spatial dynamics can be controlled by the choice of optical excitation (Sect. 4.2). The use of the above method for determining spatial dynamics to investigate the influence of scattering and plasmon coupling on the spatial dynamics of the Bloch oscillations is described (Sect. 4.3). In the final section (Sect. 4.4), we show that the harmonic Bloch oscillations are superimposed on a linear motion, which is a coherent analogue of the Shapiro effect in superconductors.

4.1 Determination of the Spatial Amplitude from Transient Peak Shifts

The basic idea of the experiment discussed here is that the Bloch-oscillating electrons, which are associated with an oscillating macroscopic dipole moment, will create a small oscillating field which is superimposed on the static bias field (see Fig. 4.1).

The modulation of the total field by the oscillating dipole moment can be detected using the Wannier–Stark ladder itself as a sensitive field sensor: the oscillation of the electric field leads to an oscillation of the transition energies. This novel technique has been proposed and used to directly resolve the spatial motion of Bloch wave packets [1, 2].

The oscillatory energy shift of the Wannier–Stark ladder transitions can be related to the relative displacement $z(t)$ of the center of mass of the electron wave function with respect to the localized hole wave function, if

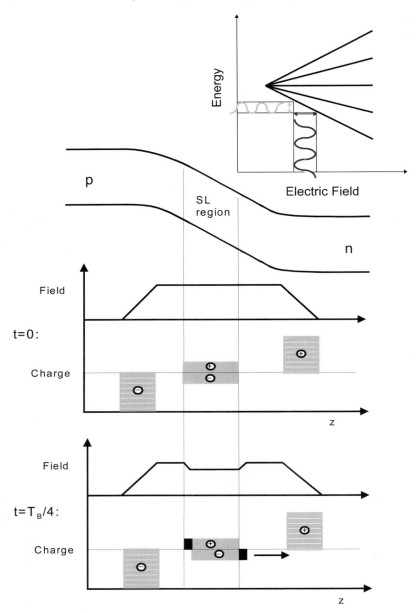

Fig. 4.1. Schematic description of the amplitude measurement. The optically generated Bloch-oscillating electrons, together with the spatially fixed holes, create an oscillating electric field (*center*, shown for delay 0 and $T_B/4$). This oscillating field modulates the Wannier–Stark ladder (*top right*); this modulation can be traced in the spectral position of the transitions

4.1 Determination of the Spatial Amplitude by Transient Peak Shifts

it is assumed that the field shift acts in a quasi-static manner. It can then be shown that [2]

$$z(t) = \frac{\epsilon_0 \epsilon_{\rm r}}{n {\rm e}^2 n_{\rm well}} \Delta E(t), \tag{4.1}$$

where $n_{\rm well}$ is the carrier density, $\epsilon_{\rm r}$ is the relative dielectric constant, and ΔE is the energy shift of a Wannier–Stark ladder transition with a Wannier–Stark ladder index n, as given in (1.11).

The only parameter needed to determine $z(t)$ is the carrier sheet density per well $n_{\rm well}$, which can, at least in principle, be determined with high precision from the laser power and the spot diameter.

It should be noted here that the quasi-static modeling of the Wannier–Stark ladder modulation needs to be refined if a very precise amplitude determination is needed. The actual modulation of the Wannier–Stark ladder is taking place at the Bloch oscillation frequency itself, and it is not obvious that the static model outlined above is a good description of the situation. A treatment of the dynamic modulation of the Wannier–Stark ladder transitions by a THz field has been performed by Dignam [3]. He predicts that the magnitude of the shifts of the Wannier–Stark ladder peaks depends strongly on the frequency of the modulation. In the low-frequency range, the static model outlined above holds; for higher frequencies, however, the amplitudes fall well below the predictions of the static model. At the Bloch frequency itself, they are enhanced again and reach about 70% of the value given by the static model. For the experiment discussed here, the errors due to this effect are smaller than the experimental error due to the uncertainty in the carrier density determination.

The dipole motion can also be detected by transmittive electro-optic sampling [4]. However, a quantitative calibration is difficult, at least if the detection is at the band gap. This is because the resonances of the electro-optic coefficients at the band gap prevent a quantitative calibration of the amplitude, since the absolute values are not well known and may also depend on the electric field, superlattice parameters, etc. Also, the electro-optic coefficients should show sign changes, which are sampled by a broad-band detection pulse in a complicated manner.

Furthermore, it was shown recently [5] that with the standard resonant electro-optic sampling setup, the weak electro-optic signal is superimposed on a third-order nonlinear (pump–probe) signal which is approximately two orders of magnitude larger. From both of these arguments, it can be expected that the resonant transmittive electro-optic sampling experiments performed in the past do not reflect the spatial dynamics of the Bloch wave packet.

Recently, electro-optic sampling experiments with nonresonant detection were performed, where the problems of resonant excitation at the band edge are not present [6]. In these experiments, the signals are about three orders of magnitude lower than for resonant excitation. However, in these experiments, the signals can be directly related to the spatial behavior of the Bloch-oscillating wave packets.

Another method which is in principle able to detect the spatial motion of the center of mass of the Bloch wave packets is a measurement of the total THz emission signal [7]. As we shall discuss in Chap. 5, the data analysis is rather involved and contains several assumptions, which limits the relevance of the results.

We now discuss in detail the experiments which detect the amplitudes of Bloch oscillations by resolving the spectral peak shifts in time. Transient four-wave-mixing experiments were carried out on a 67 Å/17 Å $GaAs/Al_{0.3}Ga_{0.7}As$ superlattice. The diffracted signal was spectrally resolved. The oscillation of the peak positions in the transient spectra is displayed in Fig. 4.2. The center transition of the Wannier–Stark ladder (hh_0), the first transition below the center (hh_{-1}), and two transitions above the center are visible. The hh_{-1} transition shows a shift to lower energy during the first part of the Bloch oscillation cycle and shifts back during the second part of the Bloch cycle, yielding an approximately harmonic shift in time, as shown in Fig. 4.3. The hh_0 center transition stays at constant energy. The hh_{+2} transition shifts first to higher energy and then to lower energy, with an amplitude which is larger than for the hh_{-1} transition.

The general behavior is thus roughly that expected from the static model assuming a single-particle Wannier–Stark ladder, although there are some

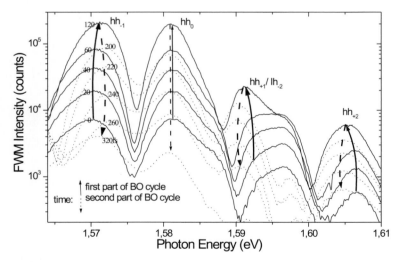

Fig. 4.2. Spectrally resolved transient four-wave-mixing data for a 67 Å/17 Å $GaAs/Al_{0.3}Ga_{0.7}As$ sample, showing the peak shifts as a function of delay time. The solid arrow indicates the peak motion fom 0 to 120 fs; the dashed arrow indicates the second part of the Bloch cycle up to 320 fs (from [2])

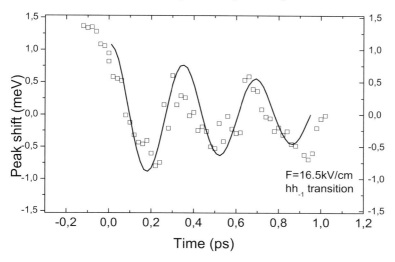

Fig. 4.3. Peak shift of the hh_{-1} peak as a function of delay time. *Symbols*: Peak shifts from experiment; *solid line*: Damped harmonic (from [2])

deviations. For instance, the hh_{+1} transition shows a quite complicated behavior owing to an anticrossing caused by a light-hole transition. In the single-particle Wannier–Stark ladder, this is not expected. However, later experiments [8] have shown that even in an anticrossing of different Wannier–Stark ladder transitions, where the peak positions do not follow the field linearly, the peaks trace the changes of the field due to the Bloch oscillations. In addition, it is visible from Fig. 4.2 that the peak shift of the hh_{+2} transition is not twice as large as that of the hh_{-1} transition. This is caused by the fact that, owing to excitonic interactions, the slope of the hh_{+2} transition is smaller.

A first step to improve the data analysis is thus not to evaluate the field shifts according to the single-particle Wannier–Stark ladder slopes expected from (4.1), but use the slopes of the Wannier–Stark ladder as actually measured in a continuous-wave optical experiment, and convert the spectral shifts to field changes using those slopes. However, the assumption of the quasi-static model is still then made. Another point which has to be taken into account is the fact that the carrier density varies with depth owing to absorption. In the analysis of the data presented here, we have taken this into account by modeling the peak shift for an inhomogeneous field [2].

Using the quasi-static model, we have derived the displacement of the wave packet from the peak shift displayed in Fig. 4.3. In deriving the amplitude, the loss of coherent carriers due to dephasing has to be taken into account: carriers which have lost their phase relative to the Bloch-oscillating ensemble will not contribute to the field oscillations. The experimentally

58   4 Interband Optical Experiments: Spatial Dynamics

observed peak shift has to be multiplied by an exponential correction function to account for these scattered carriers. The motion of the wave packet, as displayed in Fig. 4.4, is well described by an oscillation with a total amplitude of approximately 160 Å. The solid line is a harmonic function, which agrees with the experiment if one assumes an interband dephasing time of 1.2 ps, a result which is also obtained for the given experimental conditions in independent measurements [9]. The displacement as a function of time can be reasonably described by a harmonic function, as would be expected for a miniband with a harmonic dispersion (1.10). However, it is not possible to draw quantitative conclusions about the degree of harmonicity because theory has shown that the Wannier–Stark ladder can exhibit anharmonic shifts even if the driving field is harmonic [3].

The total amplitupe of the oscillation is only about twice the period of the superlattice. This comparatively small amplitude is caused by the fact that these measurements still have to be performed at rather high fields, where the wave functions are strongly localized; otherwise, the strong dephasing would prevent the completion of a Bloch oscillation cycle. Static optical experiments on superlattices have shown transitions from a localized hole to an electron wave function for up to eight wells [10]. However, for the wave function to be visible in these spectra, even a small lobe of the wave function is sufficient. For realistic laser excitation conditions, it is not possible to excite a wave packet whose center of mass is delocalized very far relative to the localized hole.

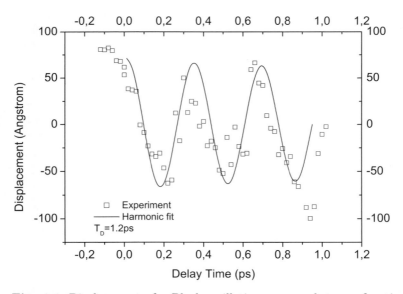

Fig. 4.4. Displacement of a Bloch-oscillating wave packet as a function of delay time. *Symbols*: Displacement from experiment; *solid line*: Harmonic fit (from [2])

4.1 Determination of the Spatial Amplitude by Transient Peak Shifts

The oscillatory peak shift observed in the experiments was superimposed on a weak linear shift. Initially, we ignored this peak shift in the data analysis, since we first associated it with the transport of incoherent carriers to the contacts. However, as we shall discuss in Sect. 4.4 in more detail, this is in fact caused by a coherent linear transport effect.

In a further experiment applying the same amplitude technique, we have measured the dependence of the Bloch oscillation amplitude on the static electric field. In the semiclassical picture, as described by (1.9), one expects the amplitude to be inversely proportional to the field.

Figure 4.5 displays the experimental data as circles. The solid line is the semiclassical result given by (1.12). The amplitude indeed decreases with electric field, as predicted by Zener. For lower fields, the amplitude data start to scatter. This is not caused by experimental error, but by the fact that in this field region, there are strong anticrossings which lead to complicated superpositions of the wave functions, leading to an erratic dependence of the amplitude on the field.

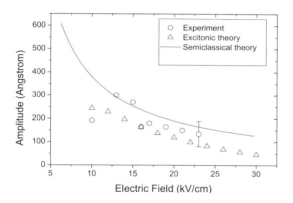

Fig. 4.5. Total amplitude of Bloch oscillations as a function of electric field. Experiment (*circles*), semiclassical theory (*solid line*), and theory including laser excitation conditions and excitonic effects (*triangles*) (from [2])

The triangles in Fig. 4.5 show a calculation using a model that takes into account excitonic effects, as presented by Dignam et al. [11], with the parameters of our experiment. The amplitudes are somewhat lower than the semiclassical limit and in good agreement with the experimental data. The main reason for the lower amplitudes predicted by the excitonic theory (in comparison with the semiclassical results) is not the electron–hole coupling (which indeed reduces the amplitude somewhat owing to the Coulomb force between electrons and holes), but the fact that the laser excitation chosen here was rather close to the center of the Wannier–Stark ladder. As will be discussed below, the full semiclassical amplitude can be expected only if the excitation is well below the center of the Wannier–Stark ladder.

4.2 Control of the Amplitude by Changing the Laser Excitation: Tuning Between Breathing Modes and Spatial Oscillations

It has already been mentioned that the displacement dynamics and the absolute amplitude of optically excited Bloch oscillations depend strongly on the composition of the wave packet created by the laser excitation, as theoretically predicted [11, 12]. This behavior can be more easily understood if one discusses it in the k-space picture. One limit is the gedankenexperiment where the initial wave packet is localized around $k = 0$. In this case all parts of the wave packet have positive mass and move initially with the field, until they reach the upper half of the band and all turn around. The same holds true if a well-localized wave packet is started at the upper edge of the band. Here, all constituents of the wave packet have negative mass, so they start against the field and oscillate. The only difference from the case of an excitation at $k = 0$ is a phase shift of the Bloch oscillation of π.

The opposite case is an experiment where the wave packet is uniformly distributed over k-space. In this case, the part with positive mass will move with the field, whereas an equally large part with negative mass will move against the field. It is obvious that the center-of-mass motion will disappear: the wave packet just oscillates between a rather localized state and a spatially distributed state, a motion which we term a *breathing mode*.

By laser excitation, one can control the wave packet dynamics and create these limiting cases, so that one can obtain a center-of-mass motion and a continuous transition to the breathing mode. Roughly speaking, the position relative to the Wannier–Stark ladder center transition (i.e. the miniband center) determines the average energy of the wave packet (see Fig. 4.6 for a schematic illustration): excitation well below the center tends to create a wave packet localized around $k = 0$, and excitation well above or at the upper edge of the band tends to create a wave packet around $k = \pi/d$. Excitation at the center of the Wannier–Stark ladder creates a wave packet spread over the band, performing a breathing-mode motion. A quantitative discussion of the dynamics following laser excitation has been given by Dignam et al. [11].

Figures 4.7 and 4.8 show calculations of the temporal dynamics of the wave packet probability density for various delay times. The cases considered are those of a single-particle Wannier–Stark ladder, with excitation of the $n = 0$ and $n = -1$ transitions, leading to an oscillating motion, and symmetric excitation of the $n = +1$, $n = 0$, and $n = -1$ transitions, leading to a breathing mode. It is obvious from the simulation that even the simple example of asymmetric excitation considered here leads to a wave packet which is quite localized in space and performs pronounced oscillations, while the width of the packet remains nearly constant. The symmetric excitation, however, leads to a wave packet which is quite localized in real space (which is expected if it is broadly distributed in k-space), but the wave packet broadens

4.2 Control of the Amplitude by Changing the Laser Excitation

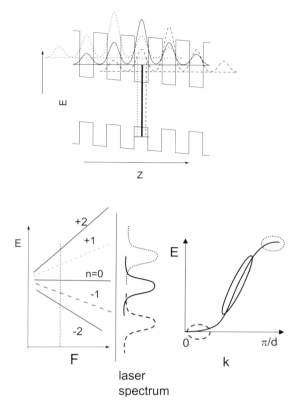

Fig. 4.6. Schematic explanation of the excitation of wave packets in real space and k-space. *Top*: superlattice structure with electron wave functions. The *solid line* marks a vertical $n = 0$ excitation, the *dashed line* a low-energy excitation, exciting the wave function in the well to the right, and the *dotted line* a high-energy excitation, exciting the well to the left. *Bottom left*: Wannier–Stark ladder and laser spectra corresponding to different excitations. *Bottom right*: k-space distributions generated by different excitations: low-energy excitation creates a wave packet localized around the lower band edge at $k = 0$, high-energy excitation creates one at the upper band edge, and excitation at the center creates one broadly distributed in k-space

with time, until it completes the oscillation cycle by coming back to a spatially localized state.

Besides the influence of the laser excitation, one also has to consider the influence of the Coulomb interaction on the spatial dynamics of the Bloch oscillations. This was again treated in detail in the paper by Dignam et al. [11]. Those authors had shown previously [13] that the linear optical spectra are influenced by the excitonic interaction, as discussed above: for instance, the spectral weights for transitions below the center increase com-

62   4 Interband Optical Experiments: Spatial Dynamics

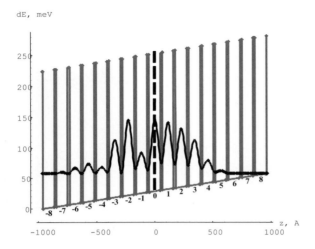

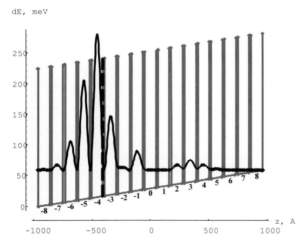

Fig. 4.7. Temporal dynamics of the wave packet probability density. The dynamics for excitation of a wave packet using the $n = 0$ and $n = -1$ transitions, leading to a harmonic spatial oscillation are shown. *Top*: delay time 0; *bottom*: delay time $T_B/4$. The *dashed vertical lines* mark the center of mass of the wave packet, which is at zero and about -400 Å amplitude, respectively.

pared with transitions above the center. In their paper about the Bloch oscillation dynamics [11], Dignam et al. showed that the general picture remains the same if Coulomb interaction is taken into account.

Figure 4.9 shows the amplitude as a function of laser energy and spectral width. The upper limit of the amplitude given by the semiclassical theory

4.2 Control of the Amplitude by Changing the Laser Excitation   63

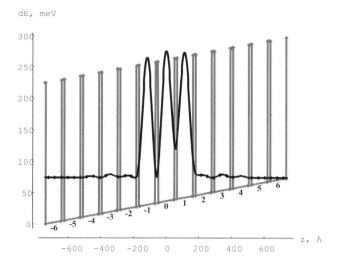

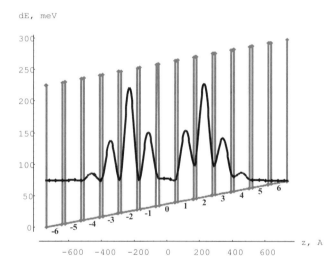

Fig. 4.8. Temporal dynamics of the wave packet probability density. The dynamics for excitation of a wave packet using the $n = +1$, $n = 0$, and $n = -1$ transitions, leading to a breathing mode, are shown. *Top*, delay time 0, *bottom*, delay time $T_B/4$. There is no motion of the center of mass for the breathing mode

(1.12) can nearly be reached if the center of the excitation is chosen well below or above the center of the Wannier–Stark ladder, which corresponds to excitation of a wave packet localized around $\bm{k} = 0$ or $\bm{k} = \pi/d$, respectively. However, for excitation close the to the center of the Wannier–Stark ladder, the amplitude of the wave packet reaches a minimum. In $\bm{k}$-space, this excitation corresponds to a wave packet which is spread over the entire Brillouin zone. In the single-particle picture, the minimum should be reached exactly for excitation at the center transition; when excitonic effects are included, theory predicts the minimum to be somewhat above. The strong reduction of

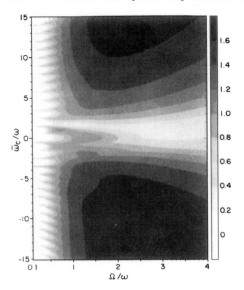

Fig. 4.9. Theoretical calculation of the amplitude of optically excited Bloch oscillations. The amplitude is shown as a grayscale (*right side*) as a function of the laser energy, expressed here in units of the Wannier–Stark ladder splitting ω, where ϖ_c is the center energy of the laser. The laser spectrum half-width Ω is again normalized by the Wannier–Stark ladder splitting. An amplitude of 2 corresponds to the semiclassical amplitude L (from [11])

the amplitude for $\Omega/\omega < 0.5$ in Fig. 4.9 is simply explained by the fact that the spectral width of the laser becomes then too small to form a wave packet. The comb-like structure is caused by the fact that a smaller spectral width can still generate a wave packet if the laser is tuned in the center between two transitions of the Wannier–Stark ladder.

The dependence of the amplitude on the laser excitation was investigated experimentally by Sudzius and coworkers [14]. These authors showed that the spatial amplitude of the oscillating wave packet can indeed be continuously tuned between symmetric breathing-mode oscillations and a harmonic spatial motion with the amplitude predicted by Zener.

The experiments were performed by tuning a smooth, Gaussian-like laser spectrum across the Wannier–Stark ladder, as shown in Fig. 4.10. The amplitude was again taken from the peak shift of the transitions of the Wannier–Stark ladder. Figure 4.11 displays the spectral shift of the hh_{-1} peak vs. delay time for various excitation conditions. The relative position of the laser excitation center wavelength is defined in units of the Wannier–Stark ladder splitting, were ω_L is the distance relative to the (experimentally observed) hh_0 transition. It is obvious that the peak shift goes through a minimum for excitation slightly below the center of the Wannier–Stark ladder. The range of excitation and the data quality are quite limited for excitation above the center: the peak shifts can then only be observed in the first Bloch oscillation cycle. This is caused simply by the fact that the relative contribution of the continuum states which are excited is much higher for excitation above the center. The free carriers which are then generated lead to a rapid dephasing of the interband coherence which is needed for the detection of the spatial motion.

4.2 Control of the Amplitude by Changing the Laser Excitation   65

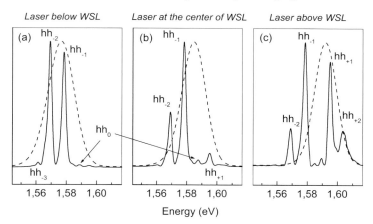

Fig. 4.10. Four-wave-mixing spectra of a 67 Å/17 Å GaAs/Al$_{0.3}$Ga$_{0.7}$As sample and laser excitation spectrum for different excitation cases in an amplitude control experiment. The vertical axis is the four-wave mixing signal in arbitrary units (from [14])

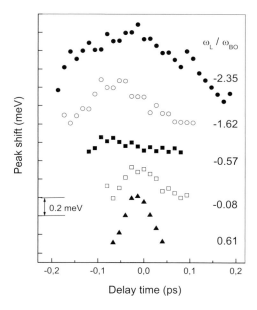

Fig. 4.11. Spectral shifts of the hh_{-1} peak during the first Bloch oscillation cycle, as a function of the laser excitation energy (given in units of the Wannier–Stark splitting, relative to the vertical $n = 0$ transition) (from [14])

Figure 4.12 shows the center-of-mass amplitude of the Bloch wave packet as a function of the excitation energy. The total amplitude of the oscillation of the wave packet is expressed in units of the Wannier–Stark ladder period (which is 84 Å in this case). The data clearly show a minimum of the amplitude for excitation near the center of the Wannier–Stark ladder. For excitation well below and above the center of the Wannier–Stark ladder,

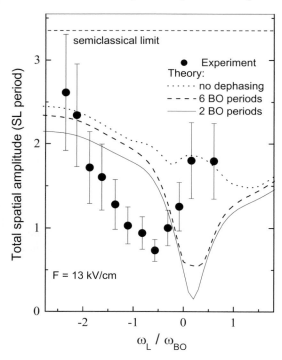

Fig. 4.12. Experimentally determined wave packet amplitude (*filled circles*) as a function of the laser excitation energy (given in units of the Wannier–Stark splitting), relative to the vertical $n = 0$ transition. The *solid, dashed and dotted lines* are results of theory for different dephasing, as described in the text (from [14])

the amplitude increases. The semiclassical amplitude given by (1.12) would be about 3.4 times the superlattice period. For excitation well below the center of the Wannier–Stark ladder, the experimental amplitude comes close to the semiclassical limit. For excitation above the center, this limit cannot be approached, for the reasons just discussed.

The experimental data were compared with theoretical results (lines in Fig. 4.12) based on a wave packet theory [11] including Coulomb interaction, as described above. The theoretical calculations predict a minimum for excitation close to the center of the Wannier–Stark ladder. However, the minimum is only pronounced if dephasing processes are taken into account (dashed and solid lines in Fig. 4.12). The simple reason is that the theory calculates the maximum amplitude which the wave packet reaches over a longer observation period. Owing to the Coulomb interaction, the symmetry between the positive-mass and negative-mass parts of the wave packet is broken: excitonic effects lead to an uneven energy spacing of the excitonic Wannier–Stark ladder levels. The exciton binding energies of different Wannier–Stark ladder transitions differ owing to the dependence of the Coulomb coupling on the wave function overlap. After a few oscillations, the two parts thus get out of phase, and an oscillation with finite amplitude results (which later on returns to the breathing-mode dynamics). This is not observed in the experiment, because the Bloch oscillation is damped long before the different parts of the wave packet get out of phase.

This effect is illustrated in more detail in Fig. 4.13, which shows the effect of the intraband dephasing on the time evolution of the dipole displacement for a wave packet excited with the laser central frequency at the $n = 0$ Wannier–Stark ladder transition. The development of a dipole amplitude with increasing delay time is clearly visible in the case where dephasing is neglected. In a real system, dephasing quickly reduces the number of coherent carriers and the nonlinear signal decays, so that the amplitude can only be derived for small delay times. The large amplitudes due to the out-of-phase motion of the constituents of the wave packet have not developed for these small delay times, and thus the low amplitudes of the initial breathing mode motion is detected in the experiment. The fact that the theoretically predicted oscillation amplitude does not reach zero for excitation close to the center (as predicted from the single-particle calculations) is also due to the influence of the Coulomb interaction: the oscillator strength of the Wannier–Stark ladder is different for transitions with positive and negative n, which prevents the creation of a fully symmetric breathing-mode wave packet with zero amplitude.

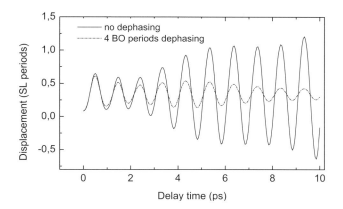

Fig. 4.13. Theoretically calculated displacement of a wave packet as a function of delay time, showing the effect of dephasing, as discussed in the text (from [14])

4.3 Influence of Scattering and Coupling to the Plasma on Bloch Oscillation Dynamics

A striking difference between experiment and theory is that theory predicts that the minimum of the amplitude is reached for excitation above the center of the Wannier–Stark ladder, whereas experiment observes a minimum amplitude clearly below the center.

Recent theoretical work [15] has shown that the previous theory could not correctly describe experiment since it did not include scattering of the Bloch-oscillating electrons. A detailed experimental study [16] of the dependence of the amplitude on the laser excitation and the carrier density has confirmed the main predictions of the extended theory [15]. In this section, we discuss the main results of these experiments and describe the comparison to the theoretical model.

First, we discuss the influence of scattering processes on the spatial dynamics of Bloch oscillations. We shall see that this influence is particularly pronounced if we investigate the control of the wave packet amplitude by changing the spectral position of the laser as discussed above. It turns out that for a full understanding of the spatial dynamics, coupling to plasmons has to be taken into account as a second effect.

The experiments were performed on the 67/17 Å $GaAs/Al_{0.3}Ga_{0.7}As$ superlattice with an electron miniband width of about 38 meV described earlier. The optical excitation was once again accomplished by means of a mode-locked Ti-sapphire laser; the detection of the wave packet motion used the amplitude measurement technique discussed above. The signals measured were the peak shifts of the spectrally resolved transient two-beam four-wave-mixing signal. The sample was held at 10 K.

Figure 4.14 shows this dependence of the field shift (which is proportional to the Bloch oscillation amplitude) for a wide range of laser excitation positions (given in units of the Wannier–Stark ladder splitting). The two sets of experimental results (circles and squares) show the amplitude for a broad and a narrow laser spectrum, respectively.

Similarly to the results discussed before, the amplitudes are large for excitation above and below the center of the Wannier–Stark ladder, and for exitation slightly below $n = 0$, a minimum is visible. As discussed above, in a single-particle picture, this minimum is expected exactly at $n = 0$; if excitonic effects are included, it is expected slightly above the center of the Wannier–Stark ladder to compensate for the asymmetric oscillator strength distribution [11]. In the following, we discuss a theoretical model which proves, in combination with experiment, that the downward shift of the minimum [14] is mainly caused by the influence of scattering on Bloch oscillations.

The theoretical model (put forward by Kosevich) is based on a semi-classical description of the dynamics of the kinetic energy and center-of-mass velocity of an electronic wave packet in a biased superlattice [15]. The semiclassical description of Bloch oscillations can be used here because of the macroscopically large number of electrons, which form a coherently oscillating wave packet.

The model is based on an effective complex electron drift velocity $\tilde{V}_z$ along the superlattice axis. The average longitudinal electron velocity V_z is determined by the real part of the effective complex velocity, $V_z = \text{Re}[\tilde{V}_z]$,

4.3 Influence of Scattering and Coupling to the Plasma

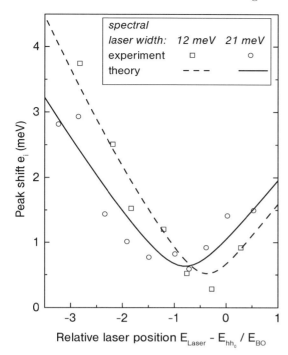

Fig. 4.14. Dependence of the peak shift of the Wannier–Stark ladder (being proportional to the Bloch oscillation amplitude) for two different widths of the laser pulse. The electric field is 15 kV/cm (from [16])

while the imaginary part of $\tilde{V}_z$ determines the average kinetic energy of the electron motion in the z-direction: $\epsilon_\parallel = -(\hbar/d)\mathrm{Im}[\tilde{V}_z]$.

From the Boltzmann equation, one can obtain the following complex equation for the average electron velocity $\tilde{V}_z$ [17]:

$$\frac{\partial \tilde{V}_z}{\partial t} + \tilde{V}_z \left(\frac{1}{\tau} + \mathrm{i}\frac{edF_z}{\hbar} \right) = \frac{e\Delta d^2}{2\hbar^2} F_z , \qquad (4.2)$$

where $\Delta d^2/(2\hbar^2)$ corresponds to the inverse reduced electron–hole effective mass m_z^{-1}, and τ is the effective intraminiband relaxation time. The complex two-component structure of (4.2) reflects the key property of Bloch oscillations: both the average group velocity and the kinetic energy of the electrons oscillate with time.

To describe the Bloch oscillations of the wave packet, (4.2) needs to be solved with the use of the semiclassical initial conditions (which are given here by the laser excitation) for the group velocity $V_z(0) = \delta_v v_z^{\max}$ and average kinetic energy $\epsilon_\parallel(0) = \delta_\epsilon \Delta/2$, where $v_z^{\max} = \Delta d/(2\hbar)$ is the maximum reduced electron–hole group velocity. The parameters δ_v and δ_ϵ stand for the dimensionless initial longitudinal reduced miniband velocity and kinetic energy of the wave packet, where $-1 \leq \delta_v \leq 1$, $0 \leq \delta_\epsilon \leq 2$. For instance, the $\delta_\epsilon = 1, \delta_v = 0$ case corresponds to the excitation of a wave packet at the center of the miniband (or at the center of the Wannier–Stark ladder in a biased superlattice). The $\delta_\epsilon < 1$ and $\delta_\epsilon > 1$ cases correspond to the excitation of a wave packet below or above the center of the miniband, respectivly.

As mentioned above, it is important to account for the interaction of Bloch oscillations with *coherent plasmons*. In a first approximation, we can also account for excitonic effects by including an effective single-particle (electron–hole) confining potential.

Both effects are parametrized by taking the electric field F_z on both sides of (4.2) as the sum of the external field F_{0z} and the induced field F_{iz}:

$$F_z = F_{0z} + F_{iz} = -eZ(4\pi N + \beta/a_B^3)/\epsilon \,, \tag{4.3}$$

where N is the electron density, $\epsilon = 12.8$ is the (static) background dielectric constant of GaAs, $a_B = \hbar^2\epsilon/me^2 = 10$ nm is the Bohr radius of an exciton, and m is the reduced electron-hole bulk effective mass. The dimensionless coefficient β, which is a parameter of a rather crude exciton model, is of the order of unity ($\beta \approx 1/7$). The induced field thus contains a term which represents the field caused by the excited carrier plasma (proportional to N) and an exciton Coulomb field.

The wave packet center-of-mass longitudinal displacement Z, which determines the induced field F_{iz}, is equal to the average relative electron–hole separation along the superlattice axis. It is related to the time-dependent wave packet velocity $V_z' = V_z - V_z^{(\infty)}$ by $\partial Z/\partial t = V_z'$, where $V_z^{(\infty)}$ is the *incoherent* steady-state longitudinal DC electron drift velocity in the superlattice: $V_z^{(\infty)} = v_z^{\max}\omega_B\tau/(1+\omega_B^2\tau^2)$ [18].

For a delay time $t \ll \tau$ after the instantaneous excitation process, (4.2) describes transient coherent Bloch oscillations of the wave packet with an *excitation-dependent* spatial amplitude $Z(t)$. The following expression for the Bloch oscillation spatial amplitude is obtained when a finite relaxation rate is included:

$$Z(t) = Z_0 + Z_B \left[\cos\phi - e^{-t/\tau^*}\cos(\omega_B^* t - \phi)\right] \,, \tag{4.4}$$

where Z_B is given by

$$Z_B = Z_B^{sc} A(\delta_\epsilon, \delta_v, \omega_B\tau) \,. \tag{4.5}$$

The amplitude factor A, which describes the deviation from the semiclassical amplitude Z_B^{sc}, is given by

$$A = \frac{\omega_B}{\sqrt{\omega_B^{*2} + 1/\tau^{*2}}}$$

$$\times \sqrt{\left(\frac{\omega_B\tau}{1+\omega_B^2\tau^2} - \delta_v\right)^2 + \left(\delta_\epsilon - \frac{\omega_B^2\tau^2}{1+\omega_B^2\tau^2}\right)^2} \,. \tag{4.6}$$

The phase factor ϕ is given by

$$\cos\phi = \frac{\omega_B}{A(\omega_B^{*2} + 1/\tau^{*2})}$$
$$\times \left[\omega_B^*\left(\frac{\omega_B^2 \tau^2}{1+\omega_B^2\tau^2} - \delta_\epsilon\right) + \frac{1}{\tau^*}\left(\delta_v - \frac{\omega_B\tau}{1+\omega_B^2\tau^2}\right)\right].$$

The results also show that the Bloch frequency is changed because of the plasma and exciton coupling described above. The resulting dynamic Bloch frequency ω_B^* is given by

$$\omega_B^* = \omega_B - \frac{e^2 d}{\epsilon \hbar}(Z_0 + Z_B \cos\phi)\left(4\pi N + \frac{\beta}{a_B^3}\right), \qquad (4.7)$$

and the dynamic relaxation rate $1/\tau^*$ is given by

$$\frac{1}{\tau^*} = \frac{1}{\tau}\left[1 + \frac{1}{\omega_B^{*2}}\frac{e^2}{\epsilon m_z}\left(4\pi N + \frac{\beta}{a_B^3}\right)\right]. \qquad (4.8)$$

Here, Z_B is the Bloch oscillation amplitude, and $Z_0 \sim 2\hbar V_z(0)/\Omega$ is the initial wave packet displacement. In case of optical excitation of a wave packet in a biased superlattice, the parameters δ_v and δ_ϵ are related by $\delta_v = \alpha(1-\delta_\epsilon)$, where $\alpha \propto \min(2\omega_B/\Omega, \Omega/2\omega_B) < 1$ and Ω is the spectral width of the pulse.

Within this model, an initial velocity $V_z(0) = v_z^{\max}\alpha(1-\delta_\epsilon)$ is given to the wave packet during a finite excitation time $2/\Omega$. For an excitation time (i.e. pulse length) shorter than the Bloch oscillation period, the initial velocity is proportional to the excitation time (the wave packet accelerates while the pulse is still on); in the opposite limit, the initial velocity is inversely proportional to the excitation time owing to the averaging over several Bloch oscillation cycles. The initial velocity $V_z(0)$ can be calculated from the experimental data using the above model: with values of $\alpha = 0.75$ and 0.47 for two laser pulse widths $\hbar\Omega = 21$ meV and 12 meV and with $1-\delta_\epsilon = 0.5$, we obtain $V_z(0) = 0.37 v_z^{\max} = 10^5$ m/s and $V_z(0) = 0.23 v_z^{\max} = 0.6 \times 10^5$ m/s, respectively.

The effective reduced mass m_z^* of the wave packet is renormalized by the wave packet excitation conditions. For $1/\tau = 0$ and $V_z(0) = 0$, it is given by $1/m_z^* = (1-\delta_\epsilon)/m_z$ [15]. Therefore, for $\delta_\epsilon \approx 1$, one observes a decrease of the wave packet spatial amplitude, which is caused by the heavy effective mass of the wave packet; this corresponds to the optical excitation of a breathing-mode wave packet with infinite average effective mass.

For a finite relaxation rate, the minimum Bloch oscillation amplitude is reached for

$$\delta_\epsilon^0 = 1 - \frac{1+\alpha\omega_B\tau}{(1+\omega_B^2\tau^2)(1+\alpha^2)}. \qquad (4.9)$$

From the results above, it follows that the spectral position and absolute value of the Bloch oscillation amplitude minimum are determined by the *bare* intraminiband relaxation rate $1/\tau$, via the parameter $\omega_B\tau > 1$. Also, the minimum of the Bloch oscillation amplitude is influenced by the laser pulse width via the parameter α [15]. In a collisionless superlattice, the minimum is exactly at the Wannier–Stark ladder center, as expected from the considerations discussed above. Since $\delta_\epsilon^0 < 1$ for a finite scattering rate, the minimum of the Bloch oscillation amplitude is always below the Wannier–Stark ladder center, because the relaxation leads to a decrease of the average wave packet kinetic energy.

The measured total peak shift e_i of the hh_{-1} transition is given by

$$e_i = \frac{2e^2 d}{\epsilon} \left(4\pi N + \frac{\beta}{a_B^3} \right) Z_B \equiv e_i^{\max} A(\delta_\epsilon, \delta_v, \omega_B \tau) \ . \tag{4.10}$$

The parameter $e_i^{\max}$ is the peak shift caused by the semiclassical Bloch oscillation amplitude for the given density of carriers N.

The lines in Fig. 4.14 display the theoretical predictions for the Bloch oscillation amplitude for two values of the spectral width of the laser pulse. By changing the spectral width of the laser pulse (i.e. the initial velocity), we can shift the spectral position of the minimum of the Bloch oscillation amplitude and the value of the amplitude at the minimum, in excellent agreement with theory.

It is predicted by (4.7) and (4.8) that the dynamic Bloch frequency ω_B^* and relaxation rate $1/\tau^*$ are shifted by both the collective (plasmon coupling) and single-particle (effective electron-hole confining potential) effects. The shift of the frequency due to the density of the carriers N was compensated in the measurements presented here by changing the bias voltage to keep the Wannier–Stark ladder splitting constant without corrections due to excitonic effects.

Figure 4.15 shows the dynamic Bloch oscillation frequency for a constant value of the Wannier–Stark ladder splitting between the transitions hh_0 and hh_{-1} as a function of the spectral position of the laser. The figure shows no dependence of the Bloch oscillation period on the carrier density. According to the theory, the change of the Bloch oscillation frequency is caused by the electron-hole confining potential and is proportional to β/a_B^3 and to the spatial amplitude $Z_B \cos \phi$ in (4.7). The observed dependence is well described by the semiclassical theory.

In a second experiment, we have investigated the *excitation density dependence* of the Bloch oscillation spatial dynamics. The excitation density dependence is of key importance for applications, since useful THz emitter devices will need high carrier densities to give reasonable output power levels.

Figure 4.16 displays the amplitude for various excitation densities (determined from photocurrent measurements). The minimum of the Bloch oscillation amplitude almost does not depend on the carrier density and temperature (from 5 to 50 K, not shown), and is therefore independent of the dynamic scattering rate $1/\tau^*$ (4.8), as predicted by theory (4.9). To model the data, we have used $\omega_B \tau = 4$, which corresponds to a bare scattering time

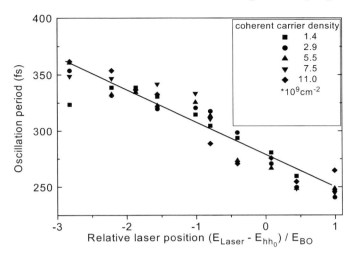

Fig. 4.15. Dynamic Bloch oscillation period as function of laser spectral position. The period changes significantly owing to the electron–hole interaction (from [16])

τ equal to 210 fs for a Bloch oscillation period of 315 fs[1]. This bare intraband damping time of the Bloch oscillations can be compared with the intraminiband relaxation time of 100 fs obtained from transport measurements at room temperature [19].

The inset in Fig. 4.16 shows the maximum depolarization energy $e_i^{\max}$, obtained from the dependence of the peak shifts on the spectral position of the laser and carrier density N. From (4.10), one can obtain the parameter $\beta \approx 0.1$ and the carrier density N from the measured dependence $e_i^{\max}(N)$. The theoretically predicted values of N agree with the density values shown in the figures. These densities were determined from the laser power and spot size.

For a larger detuning from the Wannier–Stark ladder center, the amplitude increases more strongly with detuning than it does close to the center (Fig. 4.17). The peak shift even exceeds the semiclassical value given by $e_i^{\max}(N)$ (arrows in Fig. 4.17). This is caused by interaction between Bloch oscillations and coherent plasma oscillations. This effect is well described by theory. The coupled equations for interacting Bloch and plasma oscillations in a biased superlattice can be mapped onto the excitation-dependent physical-pendulum equation [15]

$$\ddot{\theta} + \Omega_o^2 \sin\theta = 0 \,, \quad \theta = p_z d/\hbar \,, \qquad (4.11)$$

where $\Omega_o^2 = |1 - \delta_\epsilon|(\omega_{\rm pl}^2 + \omega_o^2)$, $\omega_{\rm pl} = \sqrt{4\pi N e^2/m_z \epsilon}$ is the frequency of plasma oscillations (along the superlattice axis) of the carriers with density

---

[1] The scattering time assumed here, however, is considerably shorter than the times which are derived directly from the damping of the Bloch oscillations (see Chap. 6), indicating that the model is not sufficient for a full quantitative description.

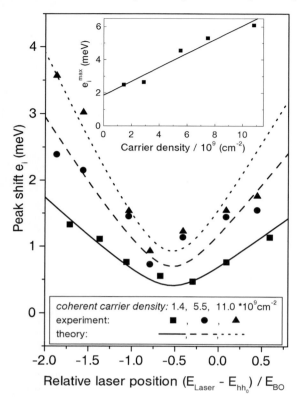

Fig. 4.16. Dependence of the Wannier–Stark ladder peak shift (which is proportional to the Bloch oscillation amplitude) on laser detuning for various densities, in the range of smaller laser detuning. The electric field is 15 kV/cm (from [16])

N, and $\omega_o = \sqrt{\beta e^2/(m_z a_B^3 \epsilon)}$ is the oscillation frequency in the effective electron–hole (single-particle) confining potential. This equation, together with the corresponding initial conditions, in general describes anharmonic oscillations.

For $1 > \delta_\epsilon$, the parameter which accounts for the anharmonicity of the oscillations is the ratio $\kappa = 2\Omega_o/\omega_B$: for $\kappa \ll 1$ the Bloch oscillations are almost sinusoidal (harmonic), while for $\kappa \approx 1$ the oscillations become strongly anharmonic and their spatial amplitude increases considerably. The frequency of the coupled Bloch and plasma oscillations is given by $\omega_B^* = \omega_B - \Omega_o^2/\omega_B$. To keep the Wannier–Stark ladder splitting, which is given by $\hbar[\omega_B - (1 - \delta_\epsilon)\omega_{pl}^2/\omega_B]$ constant, one must increase the bias (i.e. ω_B) when the density of the carriers (i.e. ω_{pl}) is increased. It can be shown that such an increase results in a monotonic increase of the parameter κ and, correspondingly, in the enhancement of the anharmonicity and amplitude of the Bloch oscillations.

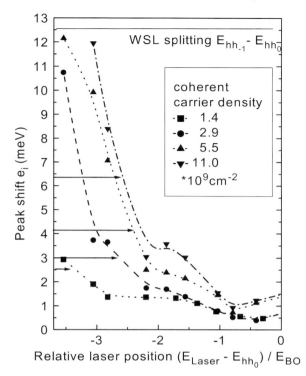

Fig. 4.17. Dependence of the Wannier–Stark ladder peak shifts (which are proportional to the Bloch oscillation amplitude) on laser detuning for various densities, in the regime of large detuning. The electric field is 15 kV/cm. The *arrows* denote the peak shifts which would correspond to the semiclassical amplitude L of Bloch oscillations. It is obvious that this limit is by far exceeded for the highest densities (from [16])

Since the frequency Ω_o in the definition of the parameter κ depends on the spectral position of the laser (via the parameter $1 - \delta_\epsilon$), the increase or decrease of κ and therefore the enhancement of the anharmonicity of the Bloch oscillations is more pronounced in the dynamics of a wave packet which is excited far below the center of the Wannier–Stark ladder (where $1-\delta_\epsilon \approx 1$). In the limit $\kappa \to 1$, the Bloch oscillations start to be destroyed by the dynamic interaction with coherent plasmons. As follows from an analysis of (4.11), in this limit the induced field F_{iz} saturates at the value of the external field $F_{0z} = \hbar\omega_B/ed$ when $e_i = 2\hbar\omega_B$, and the period of the Bloch oscillation cycle diverges (the wave packet does not return to its initial state). This conclusion from the theory agrees qualitatively with the measured peak shift and the increase in the Bloch oscillation period for high excitation densities and detunings far below the Wannier–Stark ladder center (see Fig. 4.16).

The divergence of the amplitude can also be understood in a very simple qualitative picture. If the electrons have a high density and perform Bloch oscillations with large amplitude, the field associated with the oscillations becomes so large that it exceeds the static field. In the part of the Bloch oscillation cycle when the photogenerated carriers reduce the field, the field becomes zero and the Bloch oscillations collapse. It is obvious that this effect is a natural limit on the high-density operation of a Bloch oscillator.

The results discussed here show that, although superlattices are seemingly simple systems for investigating basic quantum effects, a detailed understanding requires a rather involved analysis. It should be pointed out that the theoretical model discussed here is, despite its obvious subtleties, still a crude approximation if, for example, a description of excitonic effects is required.

4.4 Self-Induced Shapiro Effect

As discussed above, the peak shift experiments show a spectral oscillation of the Wannier–Stark ladder peaks due to the macroscopic field caused by the Bloch oscillations. Additionally, linear temporal shifts of the Wannier–Stark ladder peaks were sometimes observed. It is obvious that linear shifts could correspond to some linear motion of the electrons with the field, an effect which is not included in the simple Bloch oscillation theory. A first speculation might be that this linear shift is caused by an incoherent drift transport of carriers to the contacts. Such a drift transport is indeed expected, since not all photogenerated carriers will keep their coherence during the observation time span, and they will then be transported in a standard drift transport process. Although these incoherent carriers do not contribute to the four-wave signals anymore, the can influence the peak shifts. If one analyzes the linear shift, one obtains drift velocities which are not unreasonable.

However, as we shall discuss below, these shifts show a strong dependence on the excitation conditions, which immediately rules out the possibility that they are due to standard drift transport. As we show in the following, the linear shift is due to a coherent DC current component which is generated by the interaction of the carriers with the self-induced oscillating electric field, mediated by the free-carrier plasma which is always generated in these optical experiments. This effect, called the self-induced Shapiro effect, is a coherent analogue of the Shapiro effect observed in Josephson junctions. It turns out that this novel macroscopic quantum effect can be controlled by changing the spectral position of the exciting laser pulse, which determines the amplitude and phase of the wave packet oscillations.

4.4.1 Relation Between Bloch Oscillations and Josephson Effect

It is worthwhile to discuss here the relation between the Josephson effect, which occurs in weakly coupled superconductors, and Bloch oscillations in semiconductor superlattices. The key effect in both cases is the fact that there is a linear change in phase, which then leads to an oscillating current.

In the case of the Josephson effect, the two macroscopically coherent wave functions in the two superconductors are separated by a thin tunneling barrier. If a voltage U is applied, the energy drop across the tunneling barrier eU leads to a quantum mechanical phase $\Phi(t)$ between the two wave functions of the condensates which grows linearly with time:

$$\Phi(t) = \frac{2eU}{h} t \ . \tag{4.12}$$

Owing to the interference of the two condensates, an oscillating current

$$I(t) = I_c \sin(\Phi) \tag{4.13}$$

is generated.

This may be compared with Bloch oscillations, where the k-vector grows linearly with time (1.5). The harmonic dependence of the real-space velocity on the k-vector (1.7) then causes the oscillating Bloch current. One can also discuss Bloch oscillations, in close analogy to the Josephson effect, as an interference of the wave functions due to the potential drop eFd between adjacent wells.

In both cases, the key result is the generation of an AC effect by the application of a DC electric field to the system. The Josephson effect and Bloch oscillations show interesting extensions when both DC *and* AC voltages are applied: the current across a junction of two superconductors shows resonances (called Shapiro steps [20]), which appear at voltages which are multiples of the photon energy of the AC field, i.e.

$$eU = nh\nu \ . \tag{4.14}$$

Similar resonances in the *incoherent* (phase-independent) current in a semiconductor superlattice have been predicted by Ignatov et al. [21], for the case of a superlattice under THz illumination. The effect of the external AC field on the incoherent DC transport of the carriers is due either to photon emission (downward motion of the carriers in the Wannier–Stark ladder) or to photon absorption (upward motion), mediated by phonon emission [19]. This incoherent analogue of the Shapiro effect has been observed in superlattices by Unterrainer et al. [22], demonstrating an inverse Bloch oscillator. These experiments will be discussed in more detail in Chap. 8.

4.4.2 Observation of Linear Transport in Optically Excited Superlattices

We now discuss the linear peak shift observed in optically excited Bloch oscillations in superlattices. It can be shown by experiment that the linear

78   4 Interband Optical Experiments: Spatial Dynamics

carrier motion associated with this peak shift is not due to drift transport of *incoherent* carriers, as in the case of the THz illumination experiments just discussed, but is due to a *coherent* ensemble motion.

The experiments were performed on a 67/17 Å GaAs/Al$_{0.3}$Ga$_{0.7}$As superlattice mentioned above. The calculated electron miniband width was $\approx$ 38 meV, and the hole miniband width was $\approx$ 2 meV. The sample was held in a cryostat at 10 K. The experiments were again spectrally resolved four-wave-mixing experiments, of the type described in Sect. 4.1. The main differences between the previous experiments and those discussed here are the comparatively high excitation densities and the fact that we systematically varied the laser excitation energy.

Figure 4.18 shows the peak shift of the heavy-hole $n = -1$ Wannier–Stark ladder transition as a function of the delay time. This peak shift is proportional to the spatial motion of the photogenerated carriers. The data of Fig. 4.18 show basically three effects:

1. A large initial peak shift related to the initial buildup of the photogenerated dipole. This is caused by the fact that the photogeneration creates an electron wave packet whose center of mass is not necessarily located

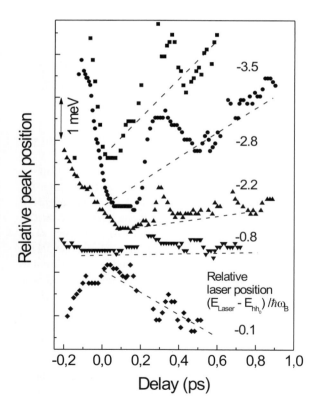

Fig. 4.18. Energies of the Wannier–Stark ladder $n = -1$ transition as a function of delay time for different laser excitation energies (from [23])

vertically above the holes. For a temporally finite laser pulse, this leads to a continuous buildup of a quasi-static dipole. The data in Fig. 4.18 show only the end of the buildup combined with the first oscillation.
2. Oscillations superimposed on the linear shift, due to Bloch oscillations of the electrons, as discussed before.
3. A linear temporal shift of the peaks, corresponding to a unidirectional motion of the carrier ensemble.

In the following, we present experiments which prove that this linear motion is not due to drift transport to the contacts of carriers which have lost their coherence. The first piece of evidence that the transport is a coherent (phase-dependent) effect is the dependence of the motion on the laser excitation conditions.

As already mentioned, the linear motion is caused by a coupling of the self-generated AC field to the excitons. As discussed in Sect. 4.2, the amplitude of this motion can be tuned from a true center-of-mass motion to a breathing mode, followed again by center-of-mass motion with inverted phase, when the center energy of the laser is shifted from below the center of the Wannier–Stark ladder to above.

It is to be expected that the linear shift of the carrier ensemble should depend on the oscillation amplitude: the shift caused by the coupling of the DC-field-induced oscillatory motion to the self-induced AC field should disappear if the oscillation amplitude is zero. This is clearly observable in Fig. 4.18: when the excitation energy is slightly below the Wannier–Stark ladder or miniband center (about 0.8 times the Wannier–Stark ladder splitting $\hbar\omega_B$), the linear shift disappears. As we have shown previously, the spatial amplitude of the oscillatory motion reaches its minimum [14, 16] exactly at this excitation position.

If the excitation photon energy is chosen to be below this minimum position, the linear shift of the $n = -1$ Wannier–Stark ladder peak is *upward in energy*, which corresponds to electron motion *downward in the Wannier–Stark ladder*, reducing the electric field. In contrast, for excitation above this minimum position, the linear shift of the $n = -1$ Wannier–Stark ladder peak is *downward in energy*, which corresponds to carrier motion *upward in the Wannier–Stark ladder*, increasing the electric field. Moving the excitation from above to below the center corresponds to a phase change of the Bloch oscillation. In the limit of excitation at the lower and upper edges of the band, this corresponds to a phase difference of π.

Figure 4.19 shows the slope of the linear carrier motion (determined from a fit to data like those shown in Fig. 4.18) as a function of the spectral position of the laser. As discussed above, the slope is close to zero when the Bloch oscillation amplitude is at its minimum, and it becomes negative as the excitation photon energy is raised above that minimum position, i.e. the direction of the coherent quasi-DC electron motion is inverted. For incoherent transport, the carriers would move along the electric field, regardless of the

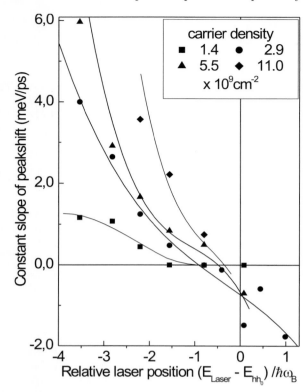

Fig. 4.19. Slopes of the Wannier–Stark ladder $n = -1$ transition shift as a function of laser excitation energy for different carrier densities. The *lines* are guides for the eye (from [23])

laser position. One might speculate that space charge effects could cause the same behavior; however, there is no reason why they should depend on the laser excitation energy.

The second piece of evidence that the motion is due to coherent quantum transport is given by the density dependence of the current: the standard carrier drift velocity, given by a balance of ballistic acceleration and momentum relaxation by scattering, should be independent of the carrier density to first order. The oscillatory field due to the macroscopic dipole of the Bloch oscillations depends linearly on the carrier density, simply because the amount of charge oscillating grows linearly with density. The overall density dependence of an incoherent Shapiro effect should thus be linear. An analysis of the experimental data indicates that the field shift increases superlinearly with increasing carrier density, which implies that the coherent quasi-DC velocity is proportional to the carrier density.

4.4.3 Theoretical Model

The self-induced Shapiro effect can be explained, in a simple picture, as follows. The Bloch-oscillating wave packet (which is composed of excitonic

Wannier–Stark ladder transitions) produces a THz field which couples to the free-carrier plasma, which is created by the continuum excitation of the Wannier–Stark ladder by the laser. The continuum performs oscillations driven by the excitonic dipole, albeit with a phase shift due to its strong damping. The phase-shifted oscillation then couples back to the excitons and allows them to be driven downwards and upwards in the field; associated with an energy loss to or gain from the free-carrier plasma. The carrier motion is shown schematically in Fig. 4.20. The exciton motion does not violate energy conservation, since it is a transient effect.

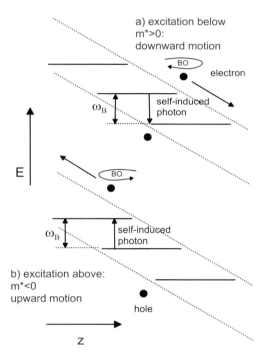

Fig. 4.20. Schematic illustration of the self-induced Shapiro effect. Depending on the effective mass of the optically generated carriers, the excitons move downwards (*top*) or upwards (*bottom*) in the Wannier–Stark ladder, i.e. photons from the plasma are emitted or absorbed, respectively

A full theoretical model of this effect needs to describe both the excitons and the free carriers, and also their coupling. A complete model which treats the whole system on an equal footing is not available at the moment. The model we discuss here describes the excitons in a state-of-the-art dynamically controlled truncation calculation of the intraband polarization, with the inclusion of a screened depolarization field [3, 15]. The plasma is described by a much simpler standard approach, as outlined below.

The time dependence of the intraband polarization P_{intra} arising from photogenerated coherent 1s excitons is calculated to second order in the optical field. The model includes intraband and interband dephasing, the full electron–hole Coulomb interaction, and the effects of the applied static electric field. The central equation describes the dynamics of the intraband correlation function $\langle B_\mu^\dagger B_\nu \rangle$ to second order in the optical field:

$$i\hbar \frac{d\langle B_\mu^\dagger B_\nu \rangle}{dt} = -\left(E_\nu - E_\mu + \frac{i\hbar}{T_{\mu\nu}}\right)\langle B_\mu^\dagger B_\nu \rangle$$
$$+ \boldsymbol{E}_{\text{opt}} \cdot \left(\boldsymbol{M}_\mu^* \langle B_\nu \rangle - \boldsymbol{M}_\nu \langle B_\mu^\dagger \rangle\right)$$
$$+ \boldsymbol{E}_{\text{AC}} \cdot \sum_\beta \left(\boldsymbol{G}_{\mu\beta}^* \langle B_\beta^\dagger B_\nu \rangle - \boldsymbol{G}_{\nu\beta} \langle B_\mu^\dagger B_\beta \rangle\right). \quad (4.15)$$

Here $B_\nu^\dagger$ is the creation operator for an exciton in the νth excitonic Wannier–Stark ladder state with energy E_ν, $\boldsymbol{M}_\nu$ is the interband dipole matrix element for the νth state, $\boldsymbol{E}_{\text{opt}}$ is the optical electric field, $\boldsymbol{G}_{\nu\beta}$ is the intraband dipole matrix element between two excitonic Wannier–Stark ladder states [3], and $T_{\mu\nu}$ is an intraband dephasing time when $\nu \neq \mu$ and an excitonic lifetime when $\nu = \mu$. For simplicity, the population decay time is taken to be infinite in all calculations below, which is a very reasonable assumption given the fact that we shall describe the temporal development in the first few picoseconds, an interval short compared with the carrier lifetime. The intraband dephasing time was set set to 1.5 ps.

The intraband polarization, to second order in the optical field, is calculated as

$$P_{\text{intra}}^{(2)} = \frac{1}{V}\sum_{\nu\mu} \boldsymbol{G}_{\nu\mu}\langle B_\nu^\dagger B_\mu \rangle, \quad (4.16)$$

were V is the volume.

The self-generated AC field appearing in (4.15) is the depolarization field of the excitons along the superlattice growth direction and is given by

$$E_{\text{AC}}(\omega) = -\frac{P_{\text{intra}}(\omega)}{\epsilon(\omega)}, \quad (4.17)$$

where $\epsilon(\omega)$ is the dielectric constant due to the *incoherent*, nonequilibrium electron–hole plasma. Once the expression for the dielectric function has been given, (4.15) – (4.17) can be solved self-consistently to yield the AC depolarization field.

Determining an accurate model for $\epsilon(\omega)$ is a formidable task. However, the essential physics can be obtained from the following simple plasma model:

$$\epsilon(\omega) = \epsilon_\infty \left(1 - \frac{\omega_p^2}{\omega^2 + i\omega/\tau}\right) \equiv R(\omega)\, e^{i\phi(\omega)}, \quad (4.18)$$

where R and ϕ are real, and $\omega_p^2 = 4\pi N e^2/m^*\epsilon_\infty$, where m^* is the reduced electron–hole average effective mass in the superlattice growth direction. The dielectric function is strongly affected by the initial energy distribution of the photogenerated incoherent electron–hole plasma. The key factor in this description is the angle ϕ, which is simply the phase shift introduced between the intraband polarization and the self-generated AC field.

From the initial conditions created by the optical pulse, it is possible to calculate the initial average effective mass for the electrons or holes by using the following expression:

$$\left\langle \frac{1}{m^*} \right\rangle = \frac{1}{\hbar^2} \int_{-\pi/d}^{\pi/d} |g(\boldsymbol{k})|^2 \, \frac{\mathrm{d}^2 E(k)}{\mathrm{d}k^2} \, \mathrm{d}k \;, \qquad (4.19)$$

where $g(\boldsymbol{k})$ is the initial distribution in $\boldsymbol{k}$-space [11]. Ignoring the electron–hole interaction, the initial average effective electron mass in the nearest-neighbor tight-binding model is given by

$$\left\langle \frac{1}{m^*} \right\rangle = \frac{1}{m_0^*} \sum_p J_{-p}(\theta) \, J_{-p-1}(\theta) \exp\left[-\left(p\omega_\mathrm{B} + E_0/\hbar - \omega_\mathrm{c}\right)^2/2\Omega^2\right]$$
$$\times \exp\left[-\left[(p+1)\omega_\mathrm{B} + E_0/\hbar + \omega_\mathrm{c}\right]^2/2\Omega^2\right] \qquad (4.20)$$
$$\div \sum_p \left| J_{-p}(\theta) \exp\left[-\left(p\omega_\mathrm{B} + E_0/\hbar - \omega_\mathrm{c}\right)^2/2\Omega^2\right]\right|^2 \;,$$

where $\theta \equiv Z_\mathrm{B}^\mathrm{sc}/d$, E_0 is the energy of the electron–hole pair in the $n = 0$ Wannier–Stark ladder state, m_0^* is the electron superlattice effective mass at the bottom of the miniband, and ω_c is the central frequency of the laser, with a Gaussian shape and a spectral width Ω.

For the parameters relevant to the system under investigation, this function is well approximated by

$$\left\langle \frac{1}{m^*} \right\rangle = \frac{-0.8}{m_0^*} \tanh\left[0.6 \left(\hbar\omega_\mathrm{c} - E_0\right)/\hbar\omega_\mathrm{B}\right] \;. \qquad (4.21)$$

Using this expression, one finds that if $\hbar\omega_\mathrm{c} < E_0$, the effective mass is positive, while if $\hbar\omega_\mathrm{c} > E_0$, the effective mass is negative. This change in the effective mass arises because for $\hbar\omega_\mathrm{c} < E_0$, the average electron energy is below the band center (where the curvature is positive), while for $\hbar\omega_\mathrm{c} < E_0$, the average energy is above the band center (where the curvature is negative).

The expression for the dielectric function then yields the result that if $m^* > 0$ then $0 < \phi < \pi$, and an applied THz field would experience loss. For $m^* < 0$, we have $-\pi < \phi < 0$, and a THz field would experience gain. It is also possible to calculate the rate at which the AC depolarization field E_AC does work on the coherent carriers:

$$\frac{\mathrm{d}W}{\mathrm{d}t} = \frac{\mathrm{d}P_\mathrm{intra}}{\mathrm{d}t} E_\mathrm{AC} \;. \qquad (4.22)$$

As a simple example, if the oscillating part of the intraband polarization is taken to be $P_\mathrm{intra}(t) = P \cos(\omega_\mathrm{B} t)$, then the time average of the rate of doing work is given by $-\omega_B P \sin[\phi(\omega_B)]/R(\omega_B)$. Thus, when $\hbar\omega_c < E_0$, the work done on the coherent carriers is negative and the electrons are driven down the Wannier–Stark ladder, while if $\hbar\omega_c > E_0$, the work done is positive and the electrons are driven up the Wannier–Stark ladder. This agrees very well with the experimentally observed motion of the carriers. It is important to note here that the upward motion observed proves the existence of *gain at THz frequencies arising from the nonequilibrium incoherent electron–hole plasma*. To our knowledge, these were the first experiments that demonstrated the existence of gain in semiconductor superlattices.

We now discuss some results from this model. Figure 4.21 displays calculated results for the self-generated AC field for three different central laser frequencies. The coherent-carrier density used in all calculations was 1.5×10^{10} cm^{-2}. To simplify the calculation, we have used the DC dielectric constant for near-DC frequencies of the polarization, and the dielectric function was evaluated at the Bloch frequency for frequencies near ω_B.

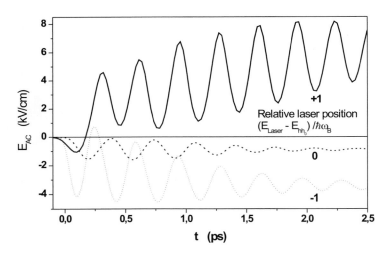

Fig. 4.21. Theoretical calculation of the Shapiro shift for different laser excitation conditions: above the center of the Wannier–Stark ladder fan chart (*solid line*), at the center (*dashed line*), and below the center (*dotted line*) (from [23])

Because the number of incoherent carriers and the precise relaxation time are not well known, we have simply set

$$\epsilon(\omega_B) = \frac{1}{2}\epsilon_\infty e^{i\pi/2} \quad \text{when} \quad \hbar\omega_c < E_0 \;,$$

$$\epsilon(\omega_B) = \frac{1}{2}\epsilon_\infty e^{-i\pi/2} \quad \text{when} \quad \hbar\omega_c > E_0 \;,$$

$$\text{and} \quad \epsilon(\omega_B) = \epsilon_\infty \quad \text{when} \quad \hbar\omega_c = E_0 \;, \tag{4.23}$$

as can be expected for a strongly damped incoherent plasma. In fact, the results are in qualitative agreement with experiment. However, to obtain quantitative agreement with the experimentally obtained slopes, one requires a carrier density two to three times the experimental value. Obviously, the simple model just outlined is not comprehensive enough to quantitatively model the coupled exciton–free-carrier plasma.

The theoretical calculations also confirm the experimental result that the slope of the linear motion is approximately proportional to the square of the density. This results simply from the fact that the rate at which work is done is proportional to the product of the AC field with the derivative of the polarization, and both of those factors are proportional to the density.

4.5 Summary

This Chapter has summarized the results of detailed experiments on the spatial dynamics of Bloch oscillations. It turns out that Bloch oscillations show a rich dynamic behavior that depends on factors such as k-space composition (which can be controlled by the laser excitation), electric field, and carrier density. It is also obvious that semiconductor superlattices, although they are a seemingly simple model system, show many intricate details, which makes a full modeling of the phenomena quite involved. A particular example is the self-induced Shapiro effect described above. However, practical devices based on Bloch oscillations will most likely be based on semiconductor superlattices, thus justifying the effort of investigating these effects in detail.

References

1. V.G. Lyssenko, G. Valusis, F. Löser, T. Hasche, K. Leo, K. Köhler, and M.M. Dignam, Proc. 23rd Int. Conf. Phys. Semicond., Berlin 1996, eds. M. Scheffler and R. Zimmermann (World Scientific, Singapore 1996), p. 1763.
2. V.G. Lyssenko, G. Valusis, F. Löser, T. Hasche, K. Leo, M.M. Dignam, and K. Köhler, Phys. Rev. Lett. **79**, 301 (1997).
3. M.M. Dignam, Phys. Rev. B **59**, 5770 (1999).
4. T. Dekorsy, P. Leisching, K. Köhler, and H. Kurz, Phys. Rev. B **50**, 8106 (1994).
5. D. Meinhold, Diploma thesis, TU Dresden (2000).
6. M. Först, G. Segschneider, T. Dekorsy, H. Kurz, and K. Köhler, Phys. Rev. B **61**, R10563 (2000).
7. R. Martini, G. Klose, H.G. Roskos, H. Kurz, H.T. Grahn, and R. Hey, Phys. Rev. B **54**, R14325 (1996).
8. F. Löser, M. Sudzius, V.G. Lyssenko, T. Hasche, K. Leo, M.M. Dignam, and K. Köhler, Phys. Stat. Sol. B **206**, 315 (1998).
9. G. Valusis, V.G. Lyssenko, D. Klatt, K.-H. Pantke, F. Löser, K. Leo, and K. Köhler, Proc. 23rd Int. Conf. Phys. Semicond., Berlin, 1996, eds. M. Scheffler and R. Zimmermann (World Scientific, Singapore 1996), p. 1783.

10. F. Agullo-Rueda, E.E. Mendez, and J.M. Hong, Phys. Rev. B **40**, 1357 (1989).
11. M. Dignam, J.E. Sipe, and J. Shah, Phys. Rev. B **49**, 10502 (1994).
12. A.M. Bouchard and M. Luban, Phys. Rev. B **47**, 6815 (1993).
13. M.M. Dignam and J.E. Sipe, Phys. Rev. Lett. **64**, 1797 (1991).
14. M. Sudzius, V.G. Lyssenko, F. Löser, K. Leo, M.M. Dignam, and K. Köhler, Phys. Rev. B **57**, 12693 (1998).
15. Yu.A. Kosevich, Phys. Rev. B **63**, 205313 (2001).
16. F. Löser, Yu.A. Kosevich, K. Köhler, and K. Leo, Phys. Rev. B **61**, R13373 (2000).
17. Yu.A. Kosevich, Ann. Phys. (Leipzig) **8**, SI-145 (1999).
18. L. Esaki and R. Tsu, IBM J. Res. Dev. **14**, 61 (1970).
19. S. Winnerl, E. Schomburg, J. Grenzer, H.-J. Regl, A.A Ignatov, A.D. Semenov, K.F. Renk, D.G. Pavel'ev, Yu. Koschurinov, B. Melzer, V. Ustinov, S. Ivanov, S. Saposchnikov, and P. Kop'ev, Phys. Rev. B **56**, 10303 (1997).
20. S. Shapiro, Phys. Rev. Lett. **11**, 80 (1963).
21. A.A. Ignatov, K.F. Renk, and E.P. Dodin, Phys. Rev. Lett. **70**, 1996 (1993).
22. K. Unterrainer, B.J. Keay, M.C. Wanke, S.J. Allen, Jr., D. Leonard, G. Medeiros-Ribeiro, U. Bhattacharya, and M.J. Rodwell, Phys. Rev. Lett. **76**, 2973 (1996).
23. F. Löser, M.M. Dignam, Yu.A. Kosevich, K. Köhler, and K. Leo, Phys. Rev. Lett. **85**, 4763 (2000).

5 Emission of Terahertz Radiation

As described by the theory set out in Chap. 1 and proven by the experiments discussed in Chap. 4, Bloch oscillations are associated with a harmonic spatial motion of carriers. It is obvious that these carriers should thus generate radiation owing to their continuous acceleration. In a semiclassical picture, the carriers perform a harmonic spatial motion, leading to dipole radiation; in a quantum electronics picture, this radiation is caused by the downward transition of carriers between states of the Wannier–Stark ladder. The radiation emitted from the carrier ensemble is coherent if the electronic coherence of the different wave packets is preserved, i.e. if the carrier ensemble performs Bloch oscillations with a common phase. Also, the radiation could be coherent if it was due to stimulated emission, i.e. in a Bloch laser.

In this chapter, we discuss experiments which have directly measured the radiation emitted from a carrier ensemble performing Bloch oscillations. As in the experiments discussed in the previous two chapters, the ensemble is coherently generated by a short laser pulse. We first briefly discuss the experimental principle and the setups used. Then, we discuss the basic observations and further experiments which have investigated the THz radiation in more detail.

5.1 Techniques of Terahertz Spectroscopy

Terahertz spectroscopy investigates transitions in the range of a few meV to a few tens of meV. The simplest approach is transmission spectroscopy, where a THz signal is passed through the sample and the change is measured. In the easiest form, this can be done using continous wave sources, for instance in a Fourier-transform infrared (FTIR) spectrometer where the intensity transmission can be determined.

Recently, more advanced time-resolved methods have been developed which measure the dynamics of the THz electric field in the time domain (see Fig. 5.1). The principle used for the generation of the THz radiation is usually the emitted transient field of accelerated carriers. The simplest approach is to irradiate a bulk GaAs or InP wafer, which emits THz radiation into free space owing to the acceleration of carriers in the surface field, which leads to a very short, spectrally broad THz pulse.

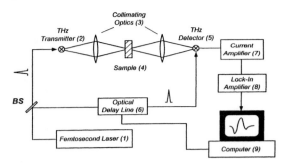

Fig. 5.1. Schematic setup of a THz time domain spectrometer (after [1])

The THz radiation is then imaged onto the sample, usually with a parabolic mirror. After transmission through the sample, the radiation is imaged, using a second parabolic mirror, onto a THz detector. Two principles have been used for the detection of the THz radiation: one of them is the Auston switch, which is, in its simplest form, a photoconductive layer where photogenerated carriers are driven by the THz field. Typically, this detector is realized in the form of a gated antenna structure.

This detector is able to time resolve the THz electric field by measuring the current driven by that field as a function of the delay of the optical pulse which creates the carriers. This is in contrast to typical measurements in the optical range, where it is not possible to resolve both the amplitude and the phase of the electric field. The spectral range which is accessible with such a gated antenna receiver is usually limited to a few THz, with the maximum sensitivity around 1 THz.

Recently, a second detection principle was found: It was demonstrated that much higher frequencies can be detected using the electro-optic effect in a thin semiconductor sample. Here, the polarization of an optical beam is rotated by the THz field. Using this method, detection at frequencies of several tens of THz has been demonstrated by [2].

The method of spectroscopy just described is limited to fairly low intensities since the power generated in the emitters has usually not exceeded the nanowatt to microwatt range and the focal diameter is large compared with those used in optical spectroscopy owing to the large wavelength. Therefore, nonlinear THz experiments cannot be performed with such a setup. Nonlinear experiments are possible using free-electron lasers (FELs). We shall discuss in Chap. 8 how such experiments have been performed on semiconductor superlattices [3].

Many recent THz experiments on semiconductors have not been performed using absorption spectroscopy, but have investigated the radiation by the sample under test itself. The experiments on Bloch oscillations which we shall discuss in the following were performed by using the biased superlattice as the emitter (see Fig. 5.2). In such experiments the sample is illuminated by a short laser pulse which generates coherently oscillating wave packets. The

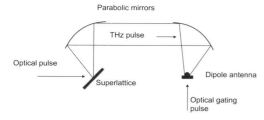

Fig. 5.2. Schematic setup of a THz emission experiment on an optically excited superlattice

THz radiation generated by these is then focused (usually by two parabolic mirrors) onto a gated antenna. The photocurrent measured across the gated antenna depends on the THz field present while the optical gating pulse is impinging on the antenna. The THz emission field of the sample can thus be measured by varying the delay time of the gating pulse relative to the optical pulse which has excited the sample.

The first such experiment which used this technique investigated the radiation from coupled double quantum wells [4]. The coupling between the two wells leads to two delocalized states. Optical excitation of a wave packet from these two states leads to a coherent oscillation between the two wells [5]. The radiation of this wave packet, which can be tuned by changing the applied voltage, was detected by the gated antenna.

5.2 Basic Experiments Observing THz Emission from Bloch Oscillations

The first experiment which detected THz radiation from Bloch-oscillating carriers was performed by Waschke et al. [6]. These authors used a 97/17 Å $Al_{0.3}Ga_{0.7}As$ superlattice held at 10 K. Owing to the large absorption of THz radiation by doped substrates, the experiments were performed in a reflection geometry as displayed in Fig. 5.2. The electric field was applied between the n-doped substrate and a semitransparent Cr front contact.

Figure 5.3 (left panel) shows the THz electric field as a function of delay time. Note that the true zero delay is around 2 ps. The signals show basically two features: a stronger signal around zero delay, and, for higher voltages, a weaker oscillatory signal. The right panel of Fig. 5.3 shows the Fourier transform of the signals. It is visible that for low fields, there is a largely field-independent contribution around 300 GHz. For higher fields, the center frequency of the fast Fourier transform (FFT) signal starts to shift linearly with the electric field. This is observed up to about 2 THz, which is at the upper edge of the frequency range of the detection setup.

As already discussed in Sect. 3.4, the static bias field is strongly influenced by the screening due to photogenerated carriers. To prove that the THz signal radiation in the Wannier–Stark field regime originates from Bloch oscillations,

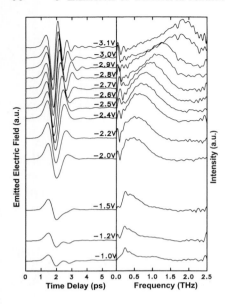

Fig. 5.3. *Left panel*: measured THz electric field vs. delay for various bias voltages. *Right panel*: Fourier transform of the THz electric field for various bias voltages (from [6])

one has to compare the THz frequencies with data for the Wannier–Stark ladder obtained under pulsed excitation.

The THz emission data of Fig. 5.3, with a correction for the spectral characteristics of the detection system, are compared with electroreflectance photocurrent data in Fig. 5.4. Owing to the field screening, the emission

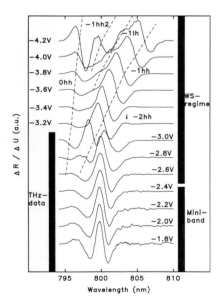

Fig. 5.4. Electroreflectance measurements of the superlattice sample under the same conditions as in the THz experiment (from [6])

frequency depends only weakly on the bias voltage up to a bias of −2.5 V. For higher voltages, the emission frequency depends linearly on the voltage. The electroreflectance spectra show the existence of a Wannier–Stark ladder in the bias range where the THz signal was observed, with a splitting which confirms that the THz emission is due to Bloch oscillations [6].

5.3 Temperature Dependence

It is interesting to study the temperature dependence of the THz emission of Bloch oscillations, since widespread application in devices will require room-temperature oscillation. The THz emission due to Bloch oscillations was studied at elevated temperatures by Waschke et al. [7].

Data for temperatures from 10 K to 120 K are shown in Fig. 5.5, which displays the THz transients as a function of temperature [7]. The emission frequency was held constant at 2 THz by adjusting the bias voltage. It is obvious that the amplitude and decay time of the THz signal decrease with increasing temperature.

Observation of the THz signal at room temperature has not possible in the experiments discussed here. This is probably caused by the fact that the damping rate increases with temperature so rapidly that Bloch oscillations can only be observed for rather high frequencies. The transmittive electro-optic sampling (TEOS) measurements of Dekorsy et al. [8] have shown that the minimum observable Bloch oscillation frequency at room temperature

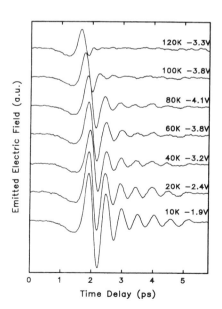

Fig. 5.5. Temperature dependence of the THz radiation signal from 10 to 120 K. The Bloch oscillation frequency was held at 2 THz by adjusting the bias voltage. The temporal shift is caused by a thermal expansion of the cold finger in the cryostat (from [7])

is about 4.5 THz, well above the frequency range addressed in these THz measurements.

The damping rate of the Bloch oscillations was analyzed using an approach discussed in detail in Sect. 6.2.1. The analysis yields a coefficient for optical-phonon scattering which predicts a room-temperature decay time of 90 fs.

5.4 Excitation Above the Band Edge

In the optical experiments on Bloch oscillations discussed so far, the excitation was at the miniband edge or slightly below the miniband edge (i.e. in the region of negative Wannier–Stark ladder index). The carriers generated in this way do not have sufficient energy to undergo significant energy relaxation processes. It is well known that the dephasing time observed in four-wave-mixing experiments depends strongly on the excitation energy: for excitation above the band edge, the dephasing time immediately becomes very short and is limited by the laser envelope. Usually, the signals become so weak in this case that they can hardly be observed. This effect can be well understood from the fact that the four-wave-mixing experiments depend on the interband, i.e. electron–hole, coherence, which drops rapidly owing to the very strong scattering of the holes when they obtain excess energy.

Interestingly, the corresponding scenario is quite different for THz emission experiments. As outlined above, the THz signal is caused not by interband coherence, but by intraband electron–electron coherence. It can be expected that the dephasing of the intraband coherence will differ from that of the interband coherence. This was indeed confirmed by THz measurements, which showed a completely different behavior when the optical excitation was changed in such a way that the excess energy of the carriers was increased [9].

Figure 5.6 shows the THz radiation from a superlattice sample with a width of the first electron miniband of 13 meV, for various excitation photon energies. Surprisingly, the oscillatory signals due to Bloch oscillations are present not only close to the band edge (photon energy 1.540 eV), but also for excitation well into the band continua. The oscillations are observed up to a photon energy of 1.642 eV, which is more than 80 meV above the band gap. For even higher excitation energies, only single-cycle transients are observed.

Theoretical calculations of the miniband electronic structure yield the following energies for the interband transitions. The transitions from the first heavy-hole miniband $hh1$ to the first electron miniband $e1$ are at 1.544–1.560 eV, and those between the second minibands $hh2$–$e2$ are at 1.624–1.675 eV. The corresponding light-hole transitions are in the energy range 1.551–1.579 eV for the $lh1$–$e1$ and 1.657–1.750 eV for the $lh2$–$e2$ transition.

5.4 Excitation Above the Band Edge   93

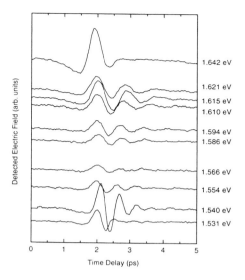

Fig. 5.6. THz transients of a sample with a width of the first electron miniband of 13 meV. Optical excitation was performed with pulses of various photon energies. The sample temperature is 10 K and the bias voltage −1.5 V (from [7])

The width of the $e2$ miniband is 45 meV. The width of the light-hole miniband is 14 meV for the $lh1$ and 45 meV for the $lh2$ miniband, respectively.

Transitions between hole and electron minibands with different indices are much weaker or entirely forbidden at the relatively low bias fields considered here and do not play a significant role. Nevertheless, such transitions can be used for probing the populations in the various minibands, as will be shown below for the detection of the $e1$ electron population via the weak $hh3$–$e1$ transition.

It is visible in Fig. 5.6 that strong oscillatory signals are detected for excitation at the band gap. For higher photon energies, the amplitude first decreases; then, it increases again to reach a second maximum when the $hh2$–$e2$ transitions are excited. The oscillations disappear below the $lh2$–$e2$ transitions.

We now address the dependence of the oscillatory signal on the electric field. This dependence is shown in Fig. 5.7 for a photon energy corresponding to the band edge excitation (1.541 eV) and in Fig. 5.8 for a photonenergy well above the band edge (1.615 eV). From the results presented above, it is known that the oscillatory signal corresponding to the band edge results from Bloch oscillations in the first miniband. Below −1.0 V, the applied field is screened by accumulation of the photogenerated charge carriers. At −1.0 V, Bloch oscillations start to be observable and can be observed up to a voltage of −2.0 V, where the Bloch frequency reaches 3 THz. At the photon energy of 1.615 eV, besides the $hh1$–$e1$ and $lh1$–$e1$ transitions, the $hh2$–$e2$ excitons and the corresponding continuum transitions are also excited. Above −1.2 V, the THz signal has an oscillatory component with a frequency that increases linearly with bias voltage.

94   5 Emission of Terahertz Radiation

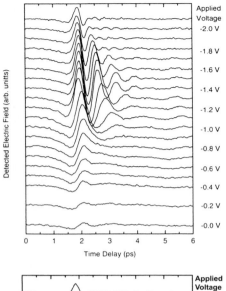

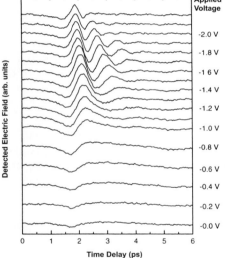

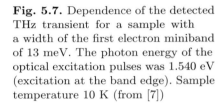

Fig. 5.7. Dependence of the detected THz transient for a sample with a width of the first electron miniband of 13 meV. The photon energy of the optical excitation pulses was 1.540 eV (excitation at the band edge). Sample temperature 10 K (from [7])

Fig. 5.8. Dependence of the detected THz transient for a sample with a width of the first electron miniband of 13 meV. The photon energy of the optical excitation pulses was 1.615 eV (excitation high above the band edge). Sample temperature 10 K (from [7])

The agreement between the oscillatory frequencies for band-gap excitation and for above-gap excitation leads directly to the conclusion that both signals result from Bloch oscillations. This conclusion holds for all excitation energies where oscillations are observed.

Similar conclusions have been drawn for another superlattice with larger miniband widths than in the 97/17 Å structure which was used to obtain the results described above. In this sample, Bloch oscillations were observed for excitation at all energies from the band gap up to approximately 1.65 eV, i.e. about three LO phonon energies above the band gap and in the regime of the $hh2$–$e2$ transitions. The Bloch oscillation origin of these signals was also demonstrated by the dependence of the oscillation frequency on the bias voltage.

Although it can be concluded from the data discussed above that the observed signals are due to Bloch oscillations even for excitation well above the band gap, the detailed nature of the relaxation processes is not immediately clear. There are basically two relaxation mechanisms possible:

- For excitation well above the lower edge of the first miniband, the radiation might be caused by continuum excitation. This would imply that the intraband coherence of the electron continuum states, which is relevant here, was surprisingly long (on the order of a picosecond or longer). It is known that the interband coherence between electrons and holes well above the band gap is on the order of a few tens of femtoseconds [10, 11]. Detailed measurements of the intraband relaxation time have not been performed. However, taking into account the fact that the LO phonon emission time is on the order of 150 fs for carriers beyond the LO phonon threshold, it is highly unlikely that the relaxation time is indeed long enough to observe Bloch oscillations from continuum carriers.
- The second possibility is that there are energy relaxation process taking place, but that they *preserve the intraband coherence* responsible for the Bloch oscillation signals. From the arguments above and the results of four-wave mixing experiments where the Bloch oscillations disappear for excitation well above the center of the first miniband, one can conclude that preservation of the intraband coherence must be possible even if the interband coherence between electrons and holes is lost.

The influence of relaxation processes on the THz emission has been investigated in detail in a Monte Carlo study [12, 13]. The theory of the authors of the study is based on the semiconductor Bloch equations in a 3D description of a semiconductor superlattice with multiple minibands. The equations include the Coulomb interaction in the time-dependent Hartree–Fock approximation and the intraband drift transport due to an applied electrical field. The coherent and incoherent (scattering) processes are described on the same kinetic level in the equations. The calculations address the low-density regime, which allows the neglect of carrier–carrier scattering. Relaxation of the carriers can proceed only via LO phonon scattering, i.e. carriers that do not reach an excess energy of 36 meV are not scattered in the simulations. The study specifically addresses the question of how relaxation of the excited electron and heavy-hole wave packets via LO phonon emission affects the phase of the intraband polarizations.

The calculations show that the decrease of the signal with increasing excitation energy is caused by destructive interference due to a scattering-induced random redistribution of carriers in k-space. The width of the electron distribution in k-space is determined by the optical excitation. When the laser is tuned to a higher energy, the width in k-space is increased, since the laser energy overlaps with a larger number of band states. The coherent transport does not change the distribution, since all carriers drift the same way in k-space according to the acceleration theorem (1.4). The maximum energy the carriers can gain is the miniband width. If the sum of the laser excess energy and the miniband width is less than the LO phonon threshold, the carriers cannot scatter in the model discussed here. In the experiment, carrier–carrier scattering and acoustic-phonon scattering will modify this. However, at low temperatures and densities, these scattering rates are comparatively small compared with LO phonon scattering.

In the four-wave-mixing experiments discussed in Chaps. 3 and 4, the excitation energies were usually chosen in such a way that most or all of the carriers stayed below this threshold; otherwise, the Bloch oscillation signal disappears very quickly. In the THz experiments discussed here, this threshold is exceeded and the carriers become redistributed in k-space. The Bloch oscillations continue to be observed if this phase-space randomization is only partial. In the theoretical calculations, the emission of even three LO phonons does not result in full randomization. Primarily, this is caused by the fact that the polar-optical LO phonon scattering prefers small momentum transfer processes.

Also, the randomization is suppressed by phase-space effects which limit the scattering channels. For instance, when the LO phonon scattering threshold has only just been passed (excitation at a photon energy of 1.580 eV, about 40 meV above the band gap), the carriers can only be scattered down to the edge of the first miniband. The distribution then starts to perform Bloch oscillations, but with a small temporal delay (of the order of the LO phonon scattering time). For excitation energies in excess of 1.620 eV, contributions from the second miniband are excited, which lead to additional terms generating THz signals.

The electrons in the second miniband deliver the most pronounced contribution for excitation at 1.640 eV, since the electron states that the Bloch wave packet is composed of are strongly delocalized and the wave packet is rather localized in k-space. The contribution of the second-miniband electrons decays in about 500 fs since the electrons are scattered to the first miniband, where they arrive, however, with a broad distribution in k-space.

We shall discuss later (Chap. 10) the possible applications of the tunable THz radiation generated in these experiments. One key point that we can mention now is the fact that the ratio of the optical input to the emitted THz output is always limited, since most of the energy is needed to create

5.5 THz Experiments on Miniband Transport and Crossover to Bloch Oscillations

In a recent series of experiments, the group of Hirakawa has investigated transport in wide minibands by using the THz detection technique. This work is an application of the THz transport techniques, used earlier by Leitenstorfer et al. [14] to study velocity overshoot in bulk semiconductors, to miniband transport in superlattices.

In a first experiment, Madhavi et al. [15] studied the THz response in a 63/7 Å $Al_{0.3}Ga_{0.7}As$ intrinsic superlattice which had a width of the first mini-

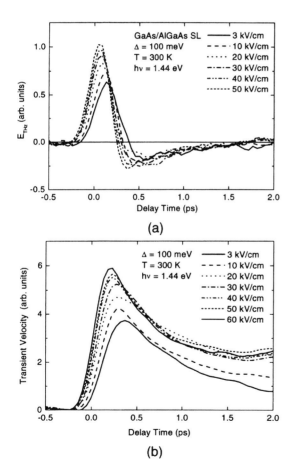

Fig. 5.9. THz transient field (**a**) and transient electron velocity (**b**) as a function of the delay time (from [15])

band of 100 meV. The sample was kept at room temperature. The THz signals were detected using an electro-optic technique with a ZnTe crystal [2, 16].

Figure 5.9a shows the THz signal as a function of the delay time for various electric fields. Initially, there is a strong positive signal due to the acceleration of the carriers in the field. This acceleration reaches a maximum, decreases again, and is then followed by a deceleration. Integration of the THz signal yields the transient velocity of the photoexcited electron ensemble directly (Fig. 5.9b). The data nicely show a pronounced velocity overshoot.

The authors show that this velocity overshoot is not caused by transitions to higher conduction band valleys, as in the classical overshoot effects observed in bulk semiconductors. The overshoot observed here is caused by a saturation and decrease of the velocity, since part of the electron ensemble reaches the upper part of the miniband, where the masses are negative. After about 2 ps, the electrons reach a steady-state velocity.

Figure 5.10 displays the steady-state velocity as a function of electric field. Initially, the velocity increases with electric field; at about 20 kV/cm, the velocity saturates or even slightly decreases. The transient photocurrent from the sample shows a similar behavior. This behavior is a transient manifestation of negative differential velocity, observed here for an accelarating ensemble in a THz experiment. The authors of [15] have also studied the acceleration behavior as a function of the *k*-space distribution of the carriers.

In a second experiment, Shimada et al. [17] have studied the THz emission of optically excited Bloch oscillation in a superlattice with a wide miniband. These authors used an 82/8 Å $Al_{0.3}Ga_{0.7}As$ intrinsic superlattice which had a width of the first miniband of 50 meV and a gap to the second miniband of 40 meV. The sample was kept at room temperature.

The time resolution was obtained by sending the input laser pulse into a Michelson interferometer to generate a double pulse. The emitted THz

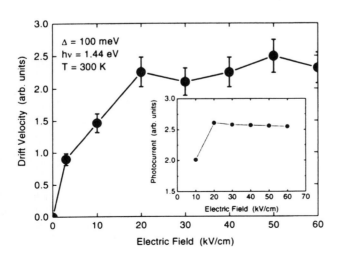

Fig. 5.10. Steady-state drift velocity as a function of bias field. The *inset* shows the photocurrent as a function of field (from [15])

5.5 THz Experiments on Miniband Transport

radiation was measured by a bolometer, as a function of the temporal delay between the two excitation pulses. In this way, a quasi-autocorrelation of the THz signals was performed.

Figure 5.11 shows the THz autocorrelation signal as a function of electric field. The signal shows a noisy peak around $\tau = 0$ due to an interference of the excitation laser pulses. For low fields, the signal shows a step. Only for higher fields does an oscillation develop.

Figure 5.12 shows the total THz signal (circles) as a function of electric field. For lower fields, the signal increases, as can be expected for miniband velocity overshoot in an increasing field. For higher fields, the signal starts to decrease again. This happens at the same field at which the pronounced oscillations start to be visible in the THz autocorrelation signal.

Figure 5.12 also shows the inverse delay time between the first and the second dip (triangles). For higher fields, the data follow the behavior expected for Bloch oscillations (dashed line); for lower fields, the time delay levels off.

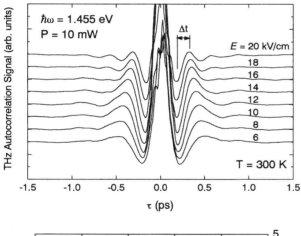

Fig. 5.11. THz autocorrelation signal recorded at room temperature (from [17])

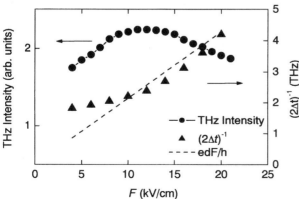

Fig. 5.12. Total THz signal (*circles*) and inverse temporal width from the first to the second dip in the THz signal (*triangles*). The *dashed line* shows the time relation expected for Bloch oscillations (from [17])

The authors interpret this behavior in terms of a crossover from miniband transport to Bloch oscillations. The comparatively wide miniband of the sample allows ballistic transport up to rather high fields. Around 10 kV/cm, however, Wannier–Stark localization sets in and Bloch oscillations become visible. The total THz signal then decreases with increasing field owing to the decreasing amplitude of the wave packet motion.

5.6 THz Emission in Crossed Electric and Magnetic Fields

Recently, Bauer et al. [18] have investigated the dynamics of optically excited wave packets in semiconductor superlattices in crossed electric and magnetic fields using THz techniques. The key observation is that there are two regimes. At low and high magnetic fields, where the Bloch oscillations are preserved, i.e. the carriers move through several Brillouin zones in k_z, where z is the growth direction of the superlattice. In addition to the oscillatory motion in the z-direction, the carriers move in the xy-plane owing to the Lorentz force, which is caused by the magnetic field perpendicular to this plane. At high magnetic fields, the Bloch oscillations become suppressed, since the carriers do not leave the first Brillouin zone anymore. In that case, the motion evolves into a cyclotron motion which has a similar spatial shape to that in the low-magnetic-field case, but a different physical origin [18], since it is basically a magnetic-field effect only.

The lateral motion of the carriers should in principle generate a Hall voltage perpendicular to the growth and magnetic-field directions (the "co-

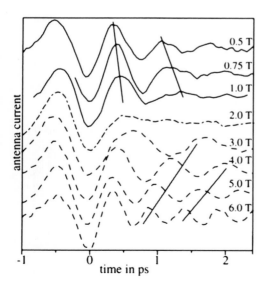

Fig. 5.13. THz signal for constant electric field and various magnetic fields. For low fields, the oscillation period increases with magnetic field; above the threshold field for the magnetic regime (2.0 T), the oscillation period decreases with magnetic field (from [18])

herent Hall effect"). However, this effect was not directly observed in the work described in [18].

Bauer et al. [18] were able to observe the transition from the low-field to the high-field regime in THz experiments (Fig. 5.13). For low fields, the period of the oscillations observed in the THz signal increases with magnetic field. At a field of 2.0 T, the transition between the low-field magneto-Bloch behavior and the high-field cyclotron regime is reached, and the oscillations become very weak. For a further increase of the field, the oscillations reappear. However, the period now decreases with increasing magnetic field. Also, the damping of the oscillations is much weaker than in the low-field case.

This behavior is summarized in Fig. 5.14, where the dependence of the oscillation frequency on the electric and magnetic fields is shown. For low magnetic fields, the period depends on both the electric and the magnetic field, as expected for a superposition of Bloch oscillations and magnetic-field motion. For the high-magnetic field region, the oscillation frequency is independent of the electric field.

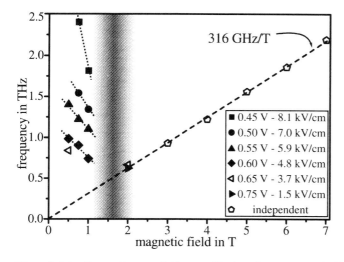

Fig. 5.14. Dependence of the oscillation frequency of the THz transients on magnetic and electric field. The gray-shaded area marks the transition between the magneto-Bloch regime (*left side*) and the cyclotron-motion regime (*right side*)(from [18])

5.7 Summary

In this chapter, we have discussed THz emission experiments on optically excited superlattices. The experiments have beautifully confirmed the existence of the "Bloch oscillator". However, despite clever attempts to improve the efficiency, it is not very likely that this type of Bloch oscillator, based on the optical generation of a wave packet, will lead to practical devices suitable for widespread use. In Chap. 10, we shall discuss different approaches to the realization of more efficient emitters.

References

1. M.C. Nuss and J. Orenstein, in *Millimeter and Submillimeter Wave Spectroscopy of Solids*, ed. G. Grüner (Springer, Berlin Heidelberg 1998), p. 7.
2. Q. Wu, M. Litz, and X.-C. Zhang, Appl. Phys. Lett. **68**, 2924 (1996); Q. Wu and X.-C. Zhang, Appl. Phys. Lett. **70**, 1784 (1997).
3. K. Unterrainer, B.J. Keay, M.C. Wanke, S.J. Allen, Jr., D. Leonard, G. Medeiros-Ribeiro, U. Bhattacharya, and M.J. Rodwell, Phys. Rev. Lett. **76**, 2973 (1996).
4. H.G. Roskos, M.C. Nuss, J. Shah, K. Leo, D.A.B. Miller, A.M. Fox, S. Schmitt-Rink, and K. Köhler, Phys. Rev. Lett. **68**, 2216 (1992).
5. K. Leo, J. Shah, E.O. Göbel, T.C. Damen, S. Schmitt-Rink, W. Schäfer, and K. Köhler, Phys. Rev. Lett. **66**, 201 (1991).
6. C. Waschke, H.G. Roskos, R. Schwedler, K. Leo, H. Kurz, and K. Köhler, Phys. Rev. Lett. **70**, 3319 (1993).
7. C. Waschke, H.G. Roskos, K. Leo, H. Kurz, and K. Köhler, Semicond. Sci. Technol. **9**, 416 (1994).
8. T. Dekorsy, R. Ott, H. Kurz, and K. Köhler, Phys. Rev. B **51**, 17275 (1995).
9. H.G. Roskos, C. Waschke, R. Schwedler, P. Leisching, Y. Dhaibi, H. Kurz, and K. Köhler, Superlatt. & Microstruct. **15**, 281 (1994).
10. P.C. Becker, H.L. Fragnito, C.H. Brito Cruz, R.L. Fork, J.E. Cunningham, J.E. Henry, and C.V. Shank, Phys. Rev. Lett. **61**, 1647 (1988).
11. W.A. Hügel, M.F. Heinrich, M. Wegener, Q.T. Vu, L. Banyai, and H. Haug, Phys. Rev. Lett. **83**, 3313 (1999).
12. T. Meier, F. Rossi, P. Thomas, and S.W. Koch, Phys. Rev. Lett. **75**, 2558 (1995).
13. F. Rossi, T. Meier, P. Thomas, S.W. Koch, P.E. Selbmann, and E. Molinari, Phys. Rev. B **51**, 16943 (1995).
14. A. Leitenstorfer, S. Hunsche, J. Shah, M.C. Nuss, and W.H. Knox, Phys. Rev. Lett. **82**, 5140 (1999).
15. S. Madhavi, M. Abe, Y. Shimada, and K. Hirakawa, Phys. Rev. B **65**, 193308 (2002).
16. A. Leitenstorfer, S. Hunsche, J. Shah, M.C. Nuss, and W.H. Knox, Appl. Phys. Lett. **74**, 1516 (1999).
17. Y. Shimada, K. Hirakawa, and S.-W. Lee, Appl. Phys. Lett. **81**, 1642 (2002).
18. T. Bauer, J. Kolb, A.B. Hummel, H.G. Roskos, Yu. Kosevich, and K. Köhler, Phys. Rev. Lett. **88**, 086801 (2002).

6 Damping of Bloch Oscillations I: Scattering

The damping mechanisms of Bloch oscillations are important both for a basic understanding of the carrier transport in superlattices and for applications. In an ideal periodic system, when only one band is considered and scattering is neglected, Bloch oscillations would last forever. In this case, the carriers would perform no net motion, i.e. there would be no net current across the system. In a real system, the coupling to phonons will interact with the oscillatory motion and lead to a superposition of quantum oscillations on linear drift transport. Additionally, as pointed out previously, coupling to higher bands (Zener tunneling) will change the simple one-band Bloch oscillation picture.

In the first chapters of this book, we have discussed several results on Bloch oscillations where damping times were determined. However, to thoroughly discuss the damping mechanisms, we give a detailed account of the damping mechanism in this and the following Chap. 7. Unavoidably, some results of the previous chapters will be mentioned again.

In the following, we shall first discuss some basics of the damping process and the methods used to investigate them. Then, in Sect. 6.2, we will discuss the damping of Bloch oscillations by scattering, either by phonons or by impurities/interface roughness. Finally, in Sect. 6.3, we shall discuss another effect of phonon coupling: if the Bloch frequency approaches the optical-phonon frequency, an interaction of Bloch oscillations and phonon oscillations takes place. We shall focus here on experiments using optical methods, but will briefly compare the results with those of transport experiments.

Damping by coupling to higher bands, i.e. Zener tunneling, will be discussed in Chap. 7.

6.1 Damping Times in Optical and Transport Experiments

In a simple picture, Bloch oscillations are a coherent process in a single band of a solid. As already mentioned in Chap. 1, this simple picture is complicated by scattering events, which break the coherence, and by coupling to other bands, which leads to complex wave packet dynamics which might mask the simple oscillation process. In terms of scattering, estimates for metals lead to the conclusion that at reasonable applied fields, the Bloch period is by far longer than the scattering time (see, e.g., [1]). Fortunately, the situation is much more benign in semiconductors.

In terms of coupling to higher bands, there was a long-standing controversy as to whether this coupling would prevent the observation of Bloch oscillations at all, or, in other words, whether the Wannier–Stark resonances exist or not. As is obvious from the results discussed in Chap. 3, experiments on semiconductor superlattices have unambiguously shown that Bloch oscillations do exist, although they are usually damped within a few oscillations. This seems to indicate that scattering by phonons or interface roughness is strong and Zener tunneling comparatively weak.

When comparing the different damping mechanisms, one has to keep in mind that Zener tunneling is a coherent process, in the sense that the coupling to higher bands is a resonant process which leads to the splitting of the Wannier–Stark states. A wave packet generated in the first band will tunnel to the higher bands, but can in principle return to the first band. In Chap. 7, we shall show this revival of the wave packet. However, such a revival is only possible if the coupling is to the second band or a few higher bands (resonant Zener tunneling). If the coupling becomes so strong that many bands are involved, a rephasing of the wave packet is not possible in a reasonable time and the Zener tunneling will lead to a complete decay of the Bloch oscillations.

In principle, the optical experiments described in Chaps. 3 and 6 and the transport experiments which will be described in Chap. 8 should yield the same results if the same physical quantities are being investigated. When comparing optical experiments with transport experiments, one has to keep in mind that optical experiments address a number of different relaxation parameters, in contrast to transport experiments, where the momentum relaxation time τ (1.3) is the central quantity: if a carrier performing Bloch oscillations undergoes a momentum relaxation process, its phase-coherent motion is terminated. It is thus obvious that Bloch oscillations can exist only if the scattering time τ is larger than the Bloch oscillation period T_B (1.10).

As already briefly discussed in Sect. 3.2, in the optical experiments one has to distinguish between the coherence of the interband, i.e. electron to hole transitions, and the coherence of the intraband, i.e. electron to electron transitions (see Fig. 3.1). Bloch oscillations are damped if the phase relation between the constituents of the wave packet, i.e. the electron Wannier–Stark ladder eigenstates, is lost. The critical parameter for this is the *intraband* damping rate $\gamma_{\text{intra}} = 1 / T_2^{\text{intra}}$, where T_2^{intra} is the intraband dephasing time. The intraband damping can be caused by transitions between states of the Wannier–Stark ladder, and by coupling of the electron Wannier–Stark ladder to a bath, e.g. the phonon system.

The interband damping rate $\gamma_{\text{inter}} = 1\,/\,T_2^{\text{inter}}$ is related to the dephasing between the photoexcited electron and hole. It has been shown that the interband and intraband dephasing times can be quite different: it is to be expected that the intraband dephasing time is longer than the interband time, since the latter contains in addition the scattering processes of the holes. Typically, hole scattering is much stronger owing to the larger hole mass.

As discussed in Chap. 3, four-wave mixing has been intensively used to investigate Bloch oscillations. In the four-wave-mixing signal, the decay of the *envelope of the four-wave-mixing signal* is related to the interband dephasing (see (3.22)). However, the damping of the harmonic modulation of the four-wave-mixing signal is not caused entirely by the interband dephasing, but is also caused by intraband dephasing: if one assumes that the difference between the interband dephasing of the two upper states is due only to intraband scattering processes between the two states (i.e. not scattering processes to a bath coupled to the three-level system), the damping of the Bloch oscillations which modulate the four-wave-mixing decay is given by the intraband dephasing only. This damping time can be obtained by a fit to the four-wave-mixing decay with an exponential modulated with a damped harmonic function.

A more direct approach is provided by pump–probe experiments, where the damping of the oscillations delivers the intraband coherence directly, at least in a simple three-level description (see (3.21)). A direct approach to the damping is also provided by THz experiments, where the decay of the radiating dipole is connected with the intraband coherence decay.

6.2 Damping of Bloch Oscillations by Phonon and Interface Scattering

The temperature dependence of the Wannier–Stark ladder was studied in detail by Mendez et al. [2], who demonstrated that in a small-period superlattice, the electronic coherence is only weakly affected by temperature. These conclusions were based on results obtained from measurements of superlattice photocurrent spectra in an external electric field.

The first experiment which addressed the damping of Bloch oscillations concentrated on the effect of the band width [3]. In particular, the Wannier–Stark ladder and Bloch oscillations were investigated in a sample where the miniband width of 62 meV was much larger than the LO phonon energy of 36 meV. The four-wave-mixing data displayed a fast decay without any modulation. The authors of [3] concluded from the experiment and a comparison with theory that strong LO phonon scattering causes the fast decay and that the critical field for observing Bloch oscillations was higher. However, a later comprehensive study by Leisching et al. [4] demonstrated Bloch oscillations in a sample which had a band width of 50 meV, even for much smaller

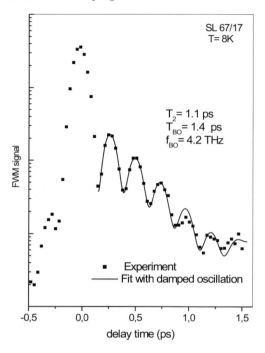

Fig. 6.1. Typical experimental four-wave-mixing trace (*symbols*) and the fit (*solid line*) to determine the interband dephasing time (from [6])

fields. These results obviously show that Bloch oscillations can exist even if the band width is much larger than the LO phonon energy. This has to be distinguished from the case where the Bloch oscillation frequency is larger than the LO frequency (see Sect. 6.3).

We now discuss a set of experiments which represent a systematic study of the damping of Bloch oscillations as a function of temperature and carrier density [5, 6]. The experiments were based on an analysis of the periodic oscillation superimposed on the four-wave-mixing data. The decay of the relative modulation of the four-wave-mixing signal was used to analyze the intraband relaxation rate γ_{intra} on the basis of (3.22). Figure 6.1 shows an example of some four-wave-mixing data together with a fit function, which is a damped harmonic modulation of the exponentially decaying four-wave-mixing envelope. The damping rate is the intraband dephasing rate γ_{intra}.

6.2.1 Decay Dynamics vs. Lattice Temperature

Figure 6.2 shows the four-wave-mixing signal of a 67 Å/17 Å $Al_{0.3}Ga_{0.7}As$ superlattice for various lattice temperatures. The spectral position of the laser and the voltage were carefully adjusted to keep the wave packet composition and the electric field constant. The fit results show that the intraband dephasing time, as derived from the decay of the harmonic modulation, is

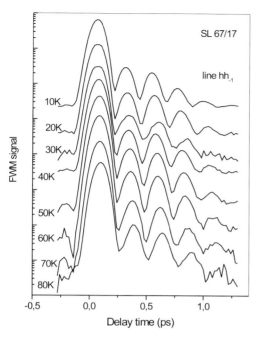

Fig. 6.2. Four-wave-mixing signals of a 67 Å/17 Å $Al_{0.3}Ga_{0.7}As$ superlattice as a function of delay time for different lattice temperatures; taken at low excitation density (from [6])

typically much longer than the interband damping time, as derived from the overall decay time of the four-wave-mixing signal. It is also visible that the overall decay of the four-wave-mixing signal becomes significantly faster with increasing temperature, while the damping of the harmonic motion is only weakly affected.

Figure 6.3 shows the intraband dephasing rate as a function of the lattice temperature [6], extracted from the data in Fig. 6.2. The results indicate that the dephasing at higher temperatures (>50 K) is due to scattering by optical phonons. At low temperatures, however, the damping is dominated by a term which is independent of both temperature and density, i.e. it is caused by neither carrier–carrier scattering nor phonon scattering. The damping as a function of density showed, in the range investigated (1×10^8 cm^{-2} to 1.5×10^9 cm^{-2}), no dependence of the intraband damping rate on density.

The experimental data have been modeled with the relation

$$\gamma_{\text{intra}} = \gamma_0 + \gamma_a T + \gamma_{\text{LO}} \exp(-\hbar\omega_{\text{LO}}/k_B T), \tag{6.1}$$

where $\gamma_0 = 0.47$ ps^{-1} is a temperature-independent constant, $\gamma_a = 3.2 \times 10^{-3}$ K^{-1} ps^{-1} is a linear coefficient modeling acoustic-phonon scattering, and $\gamma_{\text{LO}} = 50$ ps^{-1} is the prefactor of the LO phonon energy term. The LO phonon energy was assumed to be 36 meV. The result by the fit is shown as the solid line in Fig. 6.3. The value for the inhomogeneous broadening, which corresponds to an inhomogeneous line width of about 0.5 meV for a

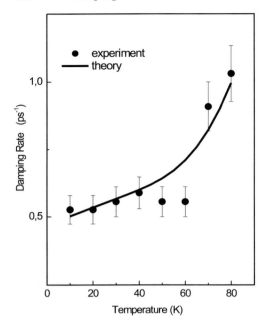

Fig. 6.3. Intraband damping rate $\gamma_{1,2}$ vs. lattice temperature (*dots*) and fit (*solid line*) as described in the text (from [6])

transition in the superlattice for high electric field, does not seem unreasonable. The acoustic-phonon damping rate of $\gamma_a = 3.2 \times 10^{-3}$ K^{-1} ps^{-1} is in reasonable agreement with measurements of the acoustic-phonon scattering rate for excitons in GaAs quantum wells [7] and corresponding theoretical results [8, 9].

An extrapolation of the damping rate to room temperature yields a damping rate of 13.8 ps^{-1}, corresponding to a damping time of ≈75 fs. Measurements of the interband dephasing in quantum wells have yielded somewhat longer dephasing times [11]. These values are in reasonable agreement with the results of Dekorsy et al. [10], who have measured a damping time of 130 fs at room temperature in a similar structure using transmittive electro-optic sampling. Dekorsy et al. have claimed that this long damping time is due to the fact that a contribution due to continuum states is measured. However, the problems associated with near-band-gap TEOS measurements, as discussed in Chap. 3, have also to be taken into account. Thus, it cannot be excluded that the results of [10] are also influenced by excitonic contributions. However, both of the above results from optical measurements are in reasonable agreement with transport measurements, which indicate a relaxation time at room temperature of about 100 fs [12].

The intraband damping rate can also be determined directly from the decay of THz Bloch oscillations as a function of temperature. The temperature-dependent data presented in [13] were analyzed using (6.1). The data were well described by the parameters $\gamma_0 = 0.08$ ps^{-1}, $\gamma_a = 0.02$ K^{-1} ps^{-1}, and $\gamma_{\text{LO}} = 20.6$ ps^{-1}. Although these data give decay rates at room temperature similar to the four-wave-mixing data just discussed, there is large disagreement in the various parameters. In particular, the THz measurements give an acoustic scattering rate which is nearly an order of magnitude larger than that measured for excitons, which is surprising since holes are expected to scatter more strongly than electrons, which are the only type of carrier involved in the THz detection. Clearly, further precise measurements over a large temperature range are needed to obtain a more complete picture.

6.2.2 Decay Dynamics vs. Excitation Intensity

We now discuss measurements of the Bloch oscillation damping time as a function of carrier density. First, we discuss again optical measurements using four-wave mixing. Here, the interband dephasing rate $\gamma_{\text{inter}} = T_2^{-1}$ and the decay rate of Bloch oscillations γ_{intra} were evaluated as a function of the carrier density by varying the laser power. Again, the wave packet composition and the electric field were kept constant by carefully adjusting the spectral position of the laser and the voltage.

The four-wave-mixing data were again evaluated by use of a damped harmonic function, as described earlier. Figure 6.4 shows the results for both the intraband (Bloch oscillations) and interband (four-wave-mixing signal) damping. It is evident that the interband dephasing rate and the decay rate of intraband polarization have quite different behaviors: the intraband decay

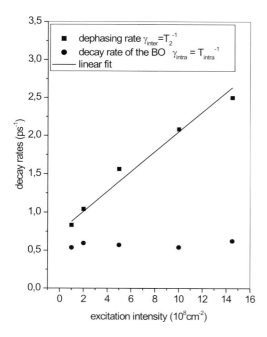

Fig. 6.4. Intraband damping rate γ_{intra} (*circles*) and interband damping rate γ_{inter} (*squares*) vs. carrier density. The linear fit (*solid line*) is described in the text (from [6])

rate γ_{intra} is nearly independent of the excitation density in the range investigated and is approximately 0.55 ps^{-1}, very similar to the value observed in the temperature-dependent measurements. In contrast, the interband dephasing rate $\gamma_{\text{inter}} = T_2^{-1}$ increases quickly. This increase can be approximated by a linear function:

$$\gamma_{\text{inter}} = \gamma_0 + \gamma_n n_{\text{exc}} , \tag{6.2}$$

where n_{exc} is the excitation density. The best fit for the coefficients yields $\gamma_0 = 0.75$ ps^{-1} and $\gamma_n = 1.3 \times 10^{-9}$ cm^2 ps^{-1}. The damping behavior can be understood as follows: The decay of the four-wave-mixing signal is caused by the sum of the electron and hole scattering rates, i.e. electron–electron, electron–hole, and hole–hole scattering. As was demonstrated in a study by Rossi et al. [14], the hole–hole interaction is the main reason for the decay of the interband polarization. With increasing intensity, the number of holes increases, resulting in more intense scattering. The density independence of the intraband damping rate suggests that electron–electron and electron–hole scattering have no important influence on the decay of the oscillations in the density range investigated.

Recent studies using the detection of THz emission [15] to study the intraband damping showed a linear increase of γ_{intra} in the density range from 2×10^9 cm^{-2} to 2×10^{10} cm^{-2} (see Fig. 6.5). The absolute values of the damping rate obtained from four-wave-mixing measurements [5] discussed

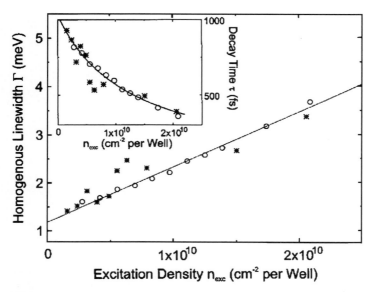

Fig. 6.5. Homogeneous line width of the THz emission from a semiconductor superlattice. Data for two excitation photon energies are shown: 1.542 eV (*stars*) and 1.557 eV (*circles*). The inset shows the decay time constants (from [15])

above and the THz results agree quite well in the density range where both experiments have taken data. One can thus conclude that the damping of the optically generated wave packets is governed by a density-independent term at low density and a linearly increasing term at higher densities. The latter term is expected for exciton–exciton scattering and has, e.g. been observed in the interband dephasing rate of quantum wells as a function of exciton density [16], for example. We shall discuss the device implications of these damping mechanisms in Chap. 10.

6.2.3 Discussion of the Decay Dynamics

We now discuss possible mechanisms which could explain the density-independent scattering term at low densities and low temperatures. The damping properties of Bloch oscillations in superlattices have been investigated theoretically by two groups [14, 17]. It was pointed out that the damping of the THz radiation due to carrier–carrier scattering is caused by the electron intraband damping [14] and is expected to be much longer than the interband damping, as has indeed been directly observed experimentally [5].

In [17], the influence of interface roughness on the damping dynamics of Bloch oscillations was investigated. It was shown that interface roughness leads to scattering, which can be described by a single, field-independent scattering time. Bloch oscillations are observed if the static field is large enough that the Bloch period is shorter than this time.

A remarkable result related to the damping of Bloch oscillations is the observation by Waschke et al. [18] that Bloch oscillations can be observed even if the interband optical excitation creates carriers in the second miniband (see Chap. 5). Since the relaxation caused by LO phonons is much faster than the Bloch oscillation period, it can be concluded that the carriers relax to the first miniband and then perform Bloch oscillations. The scattering down to the first miniband seems to preserve the intraband coherence. The interband coherence is lost in this process, as is known from the very fast disappearance of the four-wave-mixing signal once the excitation is tuned away from the band edge.

Finally, we want to briefly mention a completely different damping mechanism, which is caused by a special property of the optical experiments: The superposition of several states of the Wannier–Stark ladder which are not equidistant (owing to the Coulomb interaction) will lead to a destructive interference. It was shown in [5] that the Bloch oscillations disappear completely if many transitions of the Wannier–Stark ladder are excited simultaneously.

Figure 6.6 displays spectrally resolved four-wave-mixing signals vs. delay time for excitation of a large number of Wannier–Stark ladder states. The different traces refer to different detection energies. The traces are labeled with the transition indices, which are displayed in the inset together with the spectrum of the laser pulse used for excitation. Apparently, the oscillations

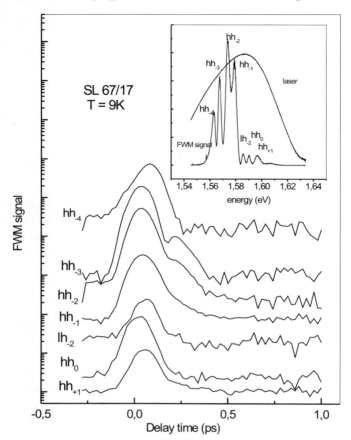

Fig. 6.6. Spectrally resolved four-wave-mixing signals from a 67 Å/17 Å $Al_{0.3}Ga_{0.7}As$ superlattice for excitation of a large number of Wannier–Stark ladder states (inset). It is obvious that the Bloch oscillations are strongly damped owing to the superposition of many transitions with different energy spacings (from [6])

are nearly completely suppressed for most of the detection energies. This effect can be explained by the interference of many nonequidistant levels of the Wannier–Stark ladder. The fact that the ladder is not equidistant is caused by the excitonic interaction: basically, the different spatial overlaps of the electron and hole for different Wannier–Stark ladder states leads to different excitonic binding energies [19]. However, one has to note that this experiment cannot completely exclude the possibility that other effects are responsible for the fast dephasing: when many levels of the Wannier–Stark ladder are excited, it cannot be avoided that a considerable number of free carriers are excited in the continuum of Wannier–Stark ladder states with lower indices.

6.3 Coupling of Bloch Oscillations and Optical Phonons

One of the key points of any device application will be the frequency range of the Bloch oscillations. For instance, the tunability range of a Bloch oscillator generating THz radiation will depend on the damping and coupling mechanisms, which limit the Bloch oscillation effect itself.

As already pointed out above, optical phonons are a strong damping mechanism for Bloch oscillations at higher temperatures. The data discussed so far have not covered the range of oscillation frequencies around or larger than the optical-phonon frequency. There is clearly the possibility that Bloch oscillations associated with a strong oscillating dipole moment will couple to polar-optical phonons when the frequency of the Bloch oscillations comes close to the phonon frequency. It is thus of great interest to investigate the behavior of Bloch oscillations in this frequency range and to investigate whether there is a coupling between Bloch oscillations and optical phonons.

An initial study [20] concluded, from electro-optic sampling data, that there is an anticrossing of the Bloch oscillations and LO phonons. However, the data were taken from a superlattice sample with a width of the first electronic miniband of 36 meV. Therefore, the regime of strong localization of the wave functions had already been reached when the Bloch oscillation frequency approached the LO phonon frequency. Thus, it was not possible to tune over the resonance.

In a later publication, Dekorsy et al. [21] investigated the coupling in more detail with different samples and came to somewhat different conclusions. They investigated two different samples with first electron miniband widths of 60 and 36 meV. The reflective electro-optic sampling (REOS) data from the sample with a first electron miniband of width 60 meV are shown in Fig. 6.7. For lower fields, the sample shows quite strongly damped Bloch oscillations.

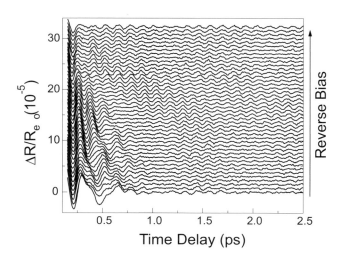

Fig. 6.7. Reflective electro-optic sampling traces of the sample with a 60 meV miniband width, taken at 10 K. (from [21])

For higher fields, the oscillation amplitude becomes much weaker but more long-lived, and reaches a damping time on the order of 10 ps.

Figure 6.8 shows Fourier transforms of the data in Fig. 6.7. Several features are visible:

- There is a bias-independent signal at the LO frequency (8.8 THz), which is attributed to the contact layers outside the superlattice part of the sample.
- There is a broad peak which shifts linearly with bias voltage (inset of the Figure), which is attributed to Bloch oscillations. For higher frequencies this peak overlaps with the phonon resonance, so that no frequency can be assigned. The linear behavior of the Bloch oscillation frequency, even close to the LO phonon energy, indicates that there is no coupled dispersion of Bloch oscillations and LO phonons, i.e. no anticrossing takes place. The upper limit of the Bloch oscillation frequency is given, in this experiment, by the laser width and is about 10 THz. However, above the LO phonon frequency, no Bloch oscillation signals are observed. This is attributed by the authors of [21] to strong damping and a weaker sensitivity of the detection mechanisms.
- A contribution is observed at the LO frequency of 8.8 THz which depends strongly on the applied bias. This contribution produces the long-lived oscillations in the time-resolved data. These oscillations are attributed to LO phonon oscillations, which are excited by coupling to Bloch oscillations.

The contribution at the LO frequency shows a quite interesting dependence on the bias voltage. Figure 6.9 displays the amplitude of the Bloch

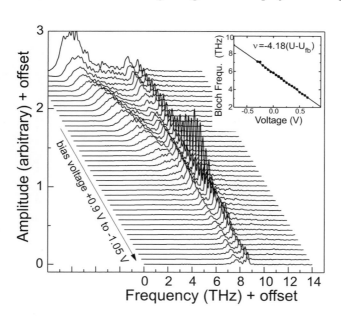

Fig. 6.8. Fourier transform of the reflective electro-optic sampling traces from Fig. 6.7. The *inset* shows the dependence of the Bloch oscillation frequency on the bias voltage (from [21])

6.3 Coupling of Bloch Oscillations and Optical Phonons

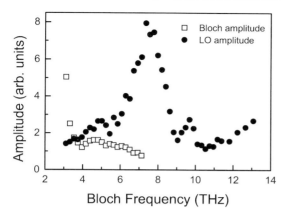

Fig. 6.9. Amplitude of the Bloch oscillations (*squares*) and of the LO phonon oscillations (*filled circles*) as a function of the Bloch frequency (from [21])

oscillation contribution and the LO phonon oscillation contribution as a function of the Bloch oscillation frequency.

The amplitude of the Bloch oscillations drops continuously with increasing frequency. The LO phonon oscillation amplitude shows a constant background from the coherent plasmon–phonon contribution from the contact layers, a pronounced resonance around 7.5 THz, and a subsequent decay towards a minimum around the LO phonon frequency. Above the LO phonon frequency a second, weaker maximum is observed. If damping were negligible, resonant driving of the LO phonons by the Bloch oscillations would occur when $\omega_B = \omega_{LO}$. However, the damping of the Bloch oscillations is strongly enhanced under this resonance condition, leading to a minimum of the LO amplitude. Above the resonance, the damping drops, leading to a second, weaker maximum in the LO amplitude.

The data of Dekorsy et al. [21] are in good agreement with the theoretical predictions of Ghosh et al. [22]. The latter authors concluded that there is no anticrossing between the Bloch oscillations and the LO phonon modes, which they ascribe to the fact that Bloch oscillations are not an elementary excitation of the system. The theory also predicts a resonant enhancement of the LO phonon amplitude when the LO phonons are in resonance with the Bloch oscillations. More recently, another theoretical study [23] concluded that an anticrossing can either occur or not occur, depending on the excitation conditions. There is still more theoretical work necessary to fully understand the theoretical implications of the Bloch oscillation–phonon coupling. It seems clear that future work needs to address this issue using a more microscopic picture.

The data of Dekorsy et al. [21] do not conclusively answer the question of whether Bloch oscillations can be observed above the LO frequency. Apparently, the damping is significant in the resonance region. Whether it becomes weak enough again at higher frequencies is an open question. According to the theory of Ghosh et al. [22], Bloch oscillations should be present at frequencies higher than the LO phonon frequency. However, this theoretical model is entirely dependent on phenomenological damping times, which need to be chosen in an ad hoc manner or taken from experiment.

6.4 Summary

The experiments on the damping of Bloch oscillations have shown that the coherence time in the conduction band, which is the key parameter that controls the damping, is determined by extrinsic effects such as interface roughness and impurities. This coherence time can be quite long at low temperatures and carrier densities, easily allowing the observation of a number of Bloch oscillation cycles. Further experiments are needed here to achieve an understanding of the relation between the damping time and the microscopic nature of the sample imperfections.

For higher temperatures and higher carrier densities, the damping is determined by phonon and carrier–carrier scattering, respectively. At room temperature, the damping time is on the order of 100–200 fs, allowing one to observe Bloch oscillations if the frequency is above a few THz. The optically measured damping times are in reasonable agreement with the damping times measured by transport experiments [12].

Experiments on the coupling of Bloch oscillations and optical phonons have explored the basic effects. For a detailed understanding, in particular whether Bloch oscillation frequencies far above the optical-phonon resonances are possible, further studies are necessary.

References

1. N.W. Ashcroft and N.D. Mermin, *Solid State Physics*, International Edition(Saunders, Philadelphia 1981), p. 225.
2. E.E. Mendez, F. Agullo-Rueda, and J.M. Hong, Appl. Phys. Lett. **56**, 2545 (1990).
3. G. von Plessen, T. Meier, J. Feldmann, E.O. Göbel, P. Thomas, K.W. Goossen, J.M Kuo, and R.F. Kopf, Phys. Rev. B **49**, 14058 (1994).
4. P. Leisching, P. Haring Bolivar, W. Beck, Y. Dhaibi, F. Brüggemann, R. Schwedler, H. Kurz, K. Leo, and K. Köhler, Phys. Rev. B **50**, 14389 (1994).
5. G. Valusis, V.G. Lyssenko, D. Klatt, K.-H. Pantke, F. Löser, K. Leo, and K. Köhler, Proc. 23rd Int. Conf. Phys. Semicond., Berlin, 1996, eds. M. Scheffler and R. Zimmermann (World Scientific, Singapore 1996), p. 1783.
6. G. Valusis, V.G. Lyssenko, F. Löser, K. Leo, Proc. 2nd Int. Conf. Excitonic Processes in Condensed Matter (EXCON '96), Kurort Gohrisch (Dresden University Press 1996), p. 115; G. Valusis, unpublished results.
7. T.C. Damen, K. Leo, J. Shah, and J.E. Cunningham, Appl. Phys. Lett. **58**, 1902 (1991).
8. J. Lee, E.S. Koteles, and M.O. Vassell, Phys. Rev. B **33**, 5512 (1985).

9. H. Stolz, D. Schwarze, W. von der Osten, and G. Weimann, Superlatt. Microstruct. **6**, 271 (1989).
10. T. Dekorsy, R. Ott, H. Kurz, and K. Köhler, Phys. Rev. B **51**, 17275 (1995).
11. K. Leo, P. Haring Bolivar, G. Maidorn, H. Kurz, and K. Köhler, Proc. 21st Int. Conf. Phys. Semicond., Beijing, 1992, eds. P. Jiang and H.-Z. Zheng (World Scientific, Singapore 1992), p. 983.
12. S. Winnerl, E. Schomburg, J. Grenzer, H.-J. Regl, A.A Ignatov, A.D. Semenov, K.F. Renk, D.G. Pavel'ev, Yu. Koschurinov, B. Melzer, V. Ustinov, S. Ivanov, S. Saposchnikov, and P. Kop'ev, Phys. Rev. B **56**, 10303 (1997).
13. C. Waschke, H.G. Roskos, K. Leo, H. Kurz, and K. Köhler, Semicond. Sci. Technol. **9**, 416 (1994).
14. F. Rossi, M. Gulia, P.E. Selbmann, E. Molinari, T. Meier, P. Thomas, and S.W. Koch, Proc. 23rd Int. Conf. Phys. Semicond., Berlin, 1996, eds. M. Scheffler and R. Zimmermann (World Scientific, Singapore 1996), p. 1775.
15. R. Martini, G. Klose, H.G. Roskos, H. Kurz, H.T. Grahn, and R. Hey, Phys. Rev. B **54**, R14325 (1996).
16. L. Schultheis, A. Honold, J. Kuhl, and C.W. Tu, Phys. Rev. B **34**, 9027 (1986).
17. E. Diez, F. Dominguez-Adame, and A. Sanchez, Microelectron. Eng. **43–44**, 117 (1998); E. Diez, R. Gomez-Alcala, F. Dominguez-Adame, A. Sanchez, and G.P. Berman, Phys. Rev. B **58**, 1146 (1998).
18. C. Waschke, H.G. Roskos, K. Leo, H. Kurz, and K. Köhler, Semicond. Sci. Technol. **9**, 416 (1994).
19. M.M. Dignam and J.E. Sipe, Phys. Rev. Lett. **64**, 1797 (1991).
20. T. Dekorsy, A.M.T. Kim, G.C. Cho, H. Kurz, and K. Köhler, Proc. 10th Int. Conf. Ultrafast Phenomena, Berlin, 1996, Springer Series in Chemical Physics, Vol. 62 (Springer, Berlin, Heidelberg 1996), p. 382.
21. T. Dekorsy, A. Bartels, H. Kurz, K. Köhler, R. Hey, and K. Ploog, Phys. Rev. Lett. **85**, 1080 (2000).
22. A.W. Ghosh, L. Jönsson, and J.W. Wilkins, Phys. Rev. Lett. **85**, 1084 (2000).
23. Yu. Kosevich, unpublished results.

7 Damping of Bloch Oscillations II: Zener Tunneling

In Sect. 2.3, we discussed various descriptions of the electronic states of a periodic potential in an electric field. The approximation most frequently used to obtain analytic solutions for the wave functions is the single-band approximation, which neglects the coupling to higher bands. Such descriptions are expected to work well at not too high fields, when the electronic states of the first band in a given well do not come into resonance with the states of the second band in a potential well close by. As discussed above, in the single-band approximation, the wave function would then completely localize in the high-field limit, and transport would be completely suppressed if there were no scattering.

However, at least in superlattices with not too weak a coupling, the resonances with adjacent wells become important at high fields. Then, the wave functions do not localize further, but again become delocalized. Therefore, starting from zero field and proceeding to high fields, there is a sequence of first delocalized, then localized, and then again delocalized wave functions.

This delocalization has also important consequences for the dynamics of carriers in the periodic potential: if the wave function were to localize completely, with increasing field the carrier transport would be characterized by Bloch oscillations with decreasing spatial amplitude, until the motion was restricted to a single well. The delocalization enables the carriers to leave their position and perform linear transport in the downward field direction.

Zener [1] first discussed this effect in 1934 and pointed out that tunneling to higher bands was a possible explanation for the electrical breakdown of semiconductors and insulators. Zener breakdown in a bulk solid is a resonant tunneling process between the valence band and conduction band [1] and occurs at high fields. Zener tunneling in bulk semiconductors was observed in transport experiments: if Zener tunneling occurs at a certain field, the conductance of the sample shows a drastic increase. The explanation of the forward current–voltage characteristics of highly doped pn diodes by Zener tunneling was given by Esaki [2] and was honored by the Nobel prize in 1973.

On the other hand, Zener breakdown in a superlattice is due to intersubband tunneling between electron miniband states. In contrast to the many recent studies on Bloch oscillations and negative differential velocity, there have been surprisingly few studies on Zener tunneling in superlattices.

The first experiment which observed Zener tunneling in a superlattice was performed by Schneider et al. [3]. The fan chart of the Wannier–Stark ladder as observed by photocurrent spectroscopy clearly showed the effects of anticrossings due to the interaction of the ladder with different minibands.

In transport experiments, the Zener effect due to resonant intersubband tunneling between different below-barrier bands was first observed by Sibille et al. [4]. The conductance of a superlattice shows peaks if different miniband states align at certain fields. In a simple picture, the transport can then be modeled by rate equations describing the process as resonant intersubband tunneling (tunneling rate R_t) and a subsequent relaxation to lower miniband states (with a relaxation rate R_{ij}). The step-like transport will increase if the coupling to a degenerate state is strong, resulting in a higher R_t.

Helm et al. [5] recently reported the observation of resonant tunneling from states of a below-barrier band to states of an above-barrier band. The effect was again shown by studying the electronic transport across the superlattice and was demonstrated by intraband spectroscopy. The experiments of Sibille et al. [4] and Helm et al. [5] will be discussed in Chap. 8.

As the results in the preceding chapters on optical experiments to investigate Bloch oscillations in superlattices have indicated, Zener tunneling affects Bloch oscillations rather weakly in standard superlattices. The data do not show any pronounced damping of the oscillations at higher fields, which would be expected if the Zener tunneling rate became comparable to the Bloch oscillation frequency. As we shall discuss below, optical spectra of standard superlattices (with several minibands below the barrier) display only weak signs of Zener tunneling.

However, as we shall discuss subsequently in this chapter, it is indeed possible to trace the Zener tunneling effect and its influence on the coherently oscillating carriers by using suitably designed samples with very high coupling to higher bands. The basic approach to designing such samples is to reduce the barrier height (by reducing the aluminum content in the barriers) so that only the first miniband is still within the potential well (shallow-well superlattices).

In the following sections, we shall discuss the implications of Zener tunneling for various phenomena:

- We shall first discuss the spectrum of a superlattice in a very high field (Sect. 7.1). Experiment and theoretical calculations show a delocalization of the Wannier–Stark ladder eigenstates, associated with an increased line broadening due to the tunneling resonances. We describe an investigation of this broadening by transient four-wave-mixing measurements and compare the results with theory.

- We shall then discuss, in Sect. 7.2, the influence of Zener tunneling on the dynamics of a wave packet composed of Wannier–Stark ladder eigenstates (i.e. the influence on the dynamics of Bloch oscillations). It turns out that the Bloch oscillations are quickly damped if strong Zener tunneling sets in.
- Finally, we address the fact that the rapid decay of the Bloch oscillations is of interferometric nature. If only one or a few states are involved, a rephasing of the wave packet (revival) is possible.

7.1 Spectrum of a Superlattice in a Very High Field

In this section, we shall first discuss the experimental observation of Zener resonances in the spectrum of semiconductor superlattices, to illustrate the basic effects. Then, we shall address theoretical considerations which outline the physics of the effects. Finally, we shall discuss studies where very detailed absorption spectra of various superlattices have been compared with experiments.

7.1.1 First Experiments

The pioneering work in this area was an experiment by Schneider et al. [3] which observed Zener tunneling in a superlattice. The authors investigated a 60-period (70 Å/9 Å) GaAs/AlAs superlattice. Low-temperature photocurrent spectroscopy was employed as experimental technique.

Figure 7.1 shows a schematic illustration of electronic resonances between nearest-neighbor wells (Fig. 7.1a) and between next-nearest-neighbor wells (Fig. 7.1b), which occur at half of this field. The optical transitions to the localized hole states are denoted by vertical arrows.

Figure 7.2 shows a theoretical calculation of the energy spectrum of the superlattice, simplified to a four-well potential. The Zener resonances between

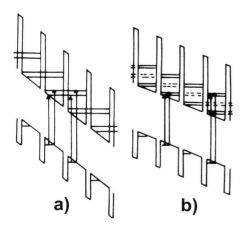

Fig. 7.1. Schematic illustration of the transitions close to a nearest neighbor (**a**) and next-nearest-neighbor (**b**) resonance (from [3])

122   7 Damping of Bloch Oscillations II: Zener Tunneling

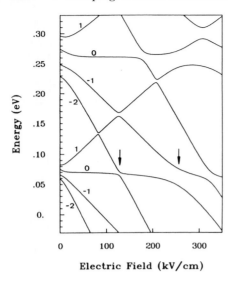

Fig. 7.2. Theoretically calculated transition spectra of the superlattice as a function of electric field, with *arrows* marking the positions of the resonances (from [3])

the first and second electronic miniband caused by the energetic degeneracy of the states (which is lifted, leading to an anticrossing) are clearly visible. For a qualitative picture, this four-well calculation can be extended to an infinite superlattice.

Figure 7.3 shows the experimentally observed fan chart of the superlattice at low temperature. The data clearly reveal the avoided crossings at Zener

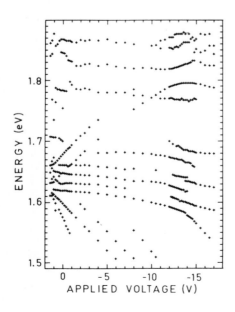

Fig. 7.3. Photocurrent spectra of the superlattice taken at low temperature (from [3])

resonances. For instance, at a voltage of about -14 V and an energy of 1.68 eV, the hh_{-1} transition of the second miniband has an anticrossing with the hh_0 transition of the first miniband.

The fact that these resonances are due to an anticrossing in the conduction miniband is confirmed by the observation that the light-hole transitions have anticrossings at the same bias field. In addition to the anticrossing of the hh_{-1} transitions, the authors of [3] also observed resonances with one and two wells in between.

7.1.2 Theoretical Considerations

As previously discussed, one-band approximations, for example the Kane model, do not describe the high-field regime correctly. With increasing field, the carrier wave functions will be less well represented by nearly symmetric, localized functions as used in the Kane approach. The Kane model fails because it leads to functions which are normalizable whereas the real wave functions are not. As discussed before, localized, normalizable functions are justified up to medium fields owing to the low tunneling probability out of the well. In the high-field limit, Kane functions will stay symmetric and localized down to an extent of about one well. This does not account for the real situation, where the wave functions start to become asymmetric.

To our knowledge, Glutsch and Bechstedt [6] showed for the first time that the localization regime is followed by a nonresonant field-induced delocalization of the electron wave function, when the field is increased. For extreme fields (where the superlattice potential $V(z)$ can be neglected), the wave function should resemble an Airy function which is delocalized. The calculations also demonstrate that for high fields, a continuous energy spectrum of the Hamiltonian is present.

Here, we discuss some numerically calculated absorption spectra of shallow-well superlattices. The calculations were performed by Glutsch and Bechstedt [7]. The main result is expressed in a one-dimensional optical density of states, which takes the form

$$D(\omega) = \int_{-\infty}^{+\infty} dE \int_{-\infty}^{+\infty} dE' \left| \int_{-\infty}^{+\infty} dz\, \psi_{eE}(z)\, \psi_{hE'}(z) \right|^2$$
$$\times \pi\, \delta\left[\hbar\omega - (E + E' + E_g) \right]. \tag{7.1}$$

The free-particle absorption, which also accounts for the in-plane motion, is proportional to the integral $\int_{-\infty}^{\omega} d\omega'\, D(\omega')$. However, the function $D(\omega)$ is better suited to comparison with experimental absorption spectra, because the maxima of D provide a good orientation for the location of the excitonic transitions (see, e.g., Fig. 4 in [7]). The joint density of states was calculated by discretization of the one-dimensional Hamiltonian. The eigenvalues and eigenvectors we evaluated by solving the characteristic polynomial of the matrix of the Hamiltonian operator. The theory considers heavy holes only. Figure 7.4 shows some results for the electron wave function, where the numerical simulation (solid line) is compared with the Kane approximation (dashed line).

In the high-field regime, owing to the well-known field-induced localization, nonvertical Wannier–Stark ladder transitions ($n \neq n'$) vanish. For further increase of the field, nondiagonal miniband transitions ($\lambda \neq \lambda'$) are observed. This underlines the fact that the selection rules found for flat band ($F = 0$) are not valid anymore. The spectrum is dominated by almost equidistant peaks, which can be assigned to direct transitions between the electron and various hole minibands. At high fields, the barrier for the electron is effectively lowered. The numerical calculation for a biased superlattice results in eigenfunctions which become asymmetric and start to delocalize (see Fig. 7.4). This effect is called *field-induced delocalization* and decreases the probability density of the electron within the well. In the delocalization regime, the asymmetric delocalized electron wave function, with a high probability density at the low-energy side of the well, should cause ($+n$) nonvertical Wannier–Stark ladder transitions to appear again in absorption. The field-induced delocalization is interpreted as nonresonant tunneling to continuum states.

In general, as proven by Avron et al. [8], the energy spectrum of a biased superlattice is continuous. In absorption, each eigenstate is weighted by the associated wave function which defines the matrix element. For states below the barrier, the Wannier–Stark ladder eigenfunctions describe sharp, spectrally equidistant resonances in the spectrum. Unconfined eigenstates above

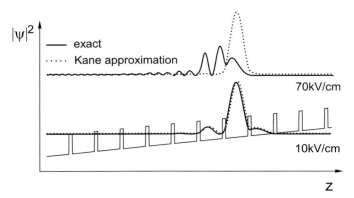

Fig. 7.4. Probability density of the electron wave function for low and high fields in a 111/17 Å superlattice: result of the numerical calculation (*solid line*); Kane approximation (*dotted line*). At 10 kV/cm, only slight deviations are observed; at 70 kV/cm, the exact wave function is completely delocalized, whereas the Kane function is completely localized in one well (from [7])

the barrier also show resonances in absorption. These states are called above-barrier states and are associated with delocalized wave functions.

For further increase of the field, the calculated joint density of states of the samples used in the work described here shows an irregular pattern of peaks. This can be interpreted as an interplay of the field-induced delocalization (non resonant tunneling to continuum states) and resonant tunneling to degenerate above-barrier states. The resonant intersubband tunneling completely dissolves the localized electron wave function, which is then extended over a few superlattice periods. If interminiband relaxation is allowed, electrical transport across the superlattice is established (for a discussion in the context of transport experiments, see Sect. 8.6.4).

7.1.3 Experimental Investigation of Zener Tunneling in Linear Optical Experiments

In the following, we discuss experiments which observe Zener tunneling in linear optical experiments. We discuss two classes of samples: the first type consists of superlattices with several minibands within the potential wells (*deep-well* samples); these were used in all of the experiments described in the previous chapters. The second type consists of specifically designed samples with low barriers which support only one miniband within the well, called *shallow-well* samples. These shallow-well samples show the Zener tunneling effects in a much more pronounced manner.

Linear Absorption of Deep-Well Superlattices First, the results of measurements on a (111 Å/17 Å) $GaAs/Ga_{0.7}Al_{0.3}As$ deep-well superlattice with a conduction band barrier height of 300 meV will be discussed. This sample structure allows two bound below-barrier electron minibands. Figure 7.5 shows a grayscale map which displays the linear absorption of the sample for fields up to 97.5 kV/cm in the spectral range of 1.51–1.67 eV. The spectra have been differentiated along the wavelength axis before the wavelength was converted into photon energy.

Figure 7.6 shows a grayscale map of the absorption according to theory, based on the model presented in [7]. Generally, experiment and theory agree well. Additional weak transitions can also be identified, which are not resolved in the experiment. It is important to note that (i) the theory does not take excitonic effects into account and (ii) it uses a large superlattice of about 100 periods, whereas the sample consists of a superlattice with 35 periods only.

At fields up to about 20 kV/cm, various strong Wannier–Stark ladder transitions evolving out of the 1D miniband can be resolved in the experiment. According to the selection rules, Wannier–Stark ladder transitions with equal miniband indices ($\lambda = \lambda'$) are allowed. Two sets of transitions are observed, which can be attributed to $(1,1)$ transitions between the first hole and the first electron miniband, centered around 1.55 eV, and $(2,2)$ transitions between

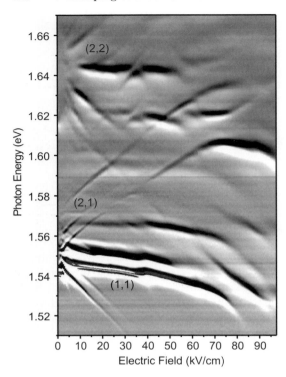

Fig. 7.5. Grayscale map for a 111 Å/17 Å $Al_{0.3}Ga_{0.7}As$ deep-well superlattice showing linear absorption spectra vs. electric field, taken at 10 K. Wannier–Stark ladder transitions between different minibands are indicated. The Zener breakdown at fields $F > 60$ kV/cm is masked by various intersubband tunneling transitions. Some transitions between minibands are denoted as (λ, λ'), where λ is the hole and λ' is the electron miniband index (from [9])

the second hole miniband and the second electron miniband, centered around 1.65 eV.

In contrast to the theory (where only heavy holes are considered), the experiment shows an additional light-hole Wannier–Stark ladder, which evolves parallel to the (1,1) Wannier–Stark ladder and is blueshifted. Additionally, the asymmetry of the Wannier–Stark ladder states due to the Coulomb potential is nicely visible. The excitonic Wannier–Stark ladder transitions with negative index have considerably higher oscillator strength than the corresponding transitions with positive index.

For fields higher than 15 kV/cm, the field-induced localization of the Wannier–Stark ladder states is visible. Nonvertical Wannier–Stark ladder transitions ($n \neq n'$) cease to exist. On the other hand, the vertical hh_0 Wannier–Stark ladder transition becomes a sharp, strong absorption peak. In this field regime, additional lines emerge, which can be attributed to nondiagonal miniband transitions ($\lambda \neq \lambda'$). These transitions occur between different hole and electron minibands. They are only weakly allowed but gain oscillator strength with increasing field. At fields of 20–60 kV/cm, the spectrum shows a set of parallel ($\lambda \neq \lambda'$) transitions. Additionally, anticrossings between different Wannier–Stark ladders can also be observed in this field regime, for example the (2,2) Wannier–Stark ladder is crossed by (2,1) transitions.

7.1 Spectrum of a Superlattice in a Very High Field   127

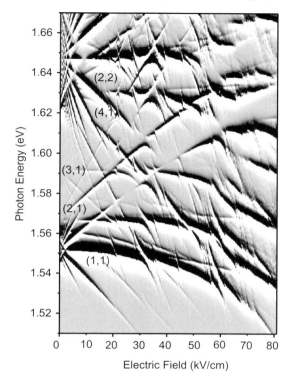

Fig. 7.6. Grayscale map of a 111 Å/17 Å $Al_{0.3}Ga_{0.7}As$ deep-well superlattice showing the joint density of states vs. electric field according to theory. The spectra have been differentiated along the energy axis. The Wannier–Stark ladder transitions between various minibands are indicated. The anticrossings at fields $F > 60$ kV/cm can be attributed to anticrossings with above-barrier states

At fields $F > 50$ kV/cm, the hh_{+1} transition of the (1,1) Wannier–Stark ladder again gains oscillator strength and can be clearly resolved. Additionally, the hh_{+1} transition of the (2,1) Wannier–Stark ladder is still present, whereas the $(-n)$ transitions have already vanished. This is caused by the field-induced delocalization of the electron wave function. The barrier is effectively lowered and the electron wave function starts to become asymmetric, tunneling nonresonantly to continuum states. This causes $(+n)$ Wannier–Stark ladder transitions to have a higher oscillator strength owing to a high probability density of the electron at the low-energy side of each well.

For fields $F > 60$ kV/cm, a distinct change in the oscillator strength and the field dependence of the transitions is demonstrated. The transitions broaden and ultimately vanish, first for the second miniband, but later also for the first miniband. The field-induced delocalization of the electron wave function causes the oscillator strength of the transitions to decrease. Furthermore, resonant tunneling to above-barrier states is observed. These states evolve from high-energy states and are degenerate with the Wannier–Stark ladder transitions at high fields $F > 60$ kV/cm.

The resonant intersubband tunneling is a clear signature of the Zener effect: the electron wave function is completely delocalized and the Wannier–

Stark ladder states break down. At this high field, the numerical results (Fig. 7.6) show an irregular pattern of absorption lines, indicating the breakdown. Additionally, a pronounced kink is observed for all transitions at a field of about 60 kV/cm, when the anticrossings occur. The states then evolve with the field at a different slope before they ultimately vanish, indicating the strong delocalization of the electron wave function. The decrease in probability density in the well is accompanied by a decrease in quantum confinement. Consequently, the field shift of the states is modified.

Besides tunneling to above-barrier states, strong anticrossings between different Wannier–Stark ladders are observed, which mask the resonant coupling to above-barrier states. Thus, the Zener breakdown in this superlattice with high barriers is attributed to an interplay between resonant intersubband tunneling to various below-barrier subbands and to above-barrier subbands.

Linear Absorption of Shallow-Well Superlattices It is obvious from the grayscale maps for deep-well superlattices (Figs. 7.5 and 7.6) that the situation is rather complicated owing to the many bands involved. Therefore, it is difficult to evaluate the details of Zener breakdown.

One way to alleviate this problem is to use superlattices with comparatively low barriers (shallow-well superlattices). This can be achieved, for example, by choosing a low Al content in the barriers of an AlGaAs superlattice. In shallow superlattices, intersubband tunneling between different below-barrier bands is avoided because only one electron miniband is present within the wells (see Fig. 7.7). Therefore, in shallow-well superlattices, Zener break-

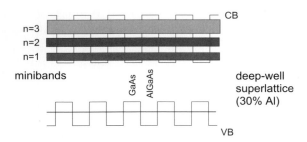

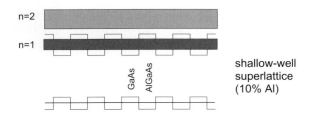

Fig. 7.7. Comparison of a *deep-well* superlattice with several minibands within the wells (*top*) and a *shallow-well* superlattice with only one miniband within the wells (*bottom*) (from [9])

down can be directly addressed as an intersubband tunneling phenomenon between bound below-barrier and delocalized above-barrier states.[1]

The absorption spectra of a 76 Å/39 Å $Al_{0.08}Ga_{0.92}As$ shallow-well superlattice are shown in Fig. 7.8. Three absorption measurements at different central frequencies (for *low*, *medium* and *high* photon energies) were carried out. The results were merged into the grayscale maps shown, which monitor the absorption over a wide spectral range.

Up to the medium-field range, the results for the shallow superlattice are similar to those for the deep 111 Å/17 Å superlattice sample. In the low-photon-energy region in Fig. 7.8, the fundamental (1, 1) Wannier–Stark ladder transitions are observed. The miniband center is at about 1.55 eV. The weakly visible fan of absorption lines centered around 1.65 eV can be

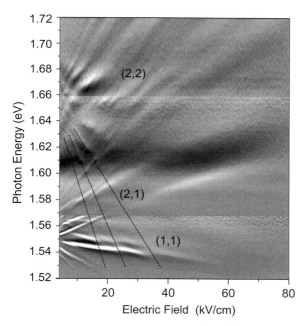

Fig. 7.8. Grayscale map for a 76 Å/39 Å $Al_{0.08}Ga_{0.92}As$ superlattice, displaying the linear absorption spectra vs. field, taken at 10 K. The data have been differentiated along the spectral axis to enhance the contrast. The *dashed lines* indicate above-barrier states, which can be traced until they cross the hh_0 (1,1) transition. The tunneling to above-barrier states results in a pronounced Zener breakdown which can be clearly observed for the hh_0 transition, which vanishes at high field (from [9])

---

[1] Strictly speaking, the states above the barriers also form bands and band gaps. However, the gaps are much smaller, so that a quasi-continuous density of states is achieved.

attributed to Wannier–Stark ladder transitions from the second heavy-hole miniband into the second, above-barrier electron band.

For fields $F > 20$ kV/cm, the field-induced localization of the electron wave function in the shallow wells is clearly visible in the experimental data. Nonvertical Wannier–Stark ladder transitions ($n \neq n'$) vanish and the direct hh_0 transition gains oscillator strength. On the other hand, nondiagonal miniband Wannier–Stark ladder transitions ($\lambda \neq \lambda'$) evolve from the lower edge of the *medium* spectral range. This weakly allowed Wannier–Stark ladder can be assigned to $(2,1)$ transitions from the second hole miniband to the first electron miniband. Strong nondiagonal Wannier–Stark ladder transitions ($n \neq n'$) are observed here. When the field is increased only $(+n)$ transitions are resolved in the experiment. This agrees with the theoretical plot. Only the $+3, +2, +1$ transitions of the $(2,1)$ Wannier–Stark ladder give pronounced peaks, even at a high electric field. Similarly to the case for the deep 111 Å/17 Å sample, this is explained by the asymmetric spatial delocalization of the electron wave function.

In the high-field regime, the hh_0 (1,1) Wannier–Stark ladder transitions are strongly modified for this shallow superlattice with 8% Al content in the barriers. Theory predicts strong coupling to above-barrier states. The experimentally observed absorption of above-barrier states is marked by dashed lines in Fig. 7.8. These states can be traced until resonant tunneling between the electron below-barrier state (observed as an hh_0 Wannier–Stark ladder transition) and above-barrier states occurs. Simultaneously, a strong broadening of the hh_0 transition is observed for $F > 35$ kV/cm. For higher fields, the oscillator strength of the hh_0 transition decreases, and ultimately vanishes at a field of about 50 kV/cm. *This feature is a clear indication of field-induced delocalization*: in the high-field regime, resonant intersubband tunneling to above-barrier states and the delocalization of the electron wave function due to nonresonant tunneling to continuum states can be clearly demonstrated. Owing to the interplay of these two effects the electron wave function is dissolved and a pronounced Zener breakdown is observed.

The theoretical plot of the joint density of states of this superlattice sample shown in Fig. 7.9 compares very well with the experiment. It also shows that the absorption for the transitions of the first miniband disappears at above 50 kV/cm. The theory also nicely shows that this is caused by strong interaction with a multitude of above-barrier states from higher minibands, which cross into the density of states of the first miniband. Due to the spatial delocalization associated with this interaction, the absorption to the transitions of the localized holes disappears.

Figure 7.10 shows absorption spectra at various fields. These spectra are cross sections of the low-field spectral range shown in Fig. 7.8. The flat-field ($F = 0$) plot shows a strong excitonic heavy-hole transition and the miniband absorption of the associated excitonic ionization continuum. There are slight modulations of the miniband absorption, as shown in the inset, for example

7.1 Spectrum of a Superlattice in a Very High Field 131

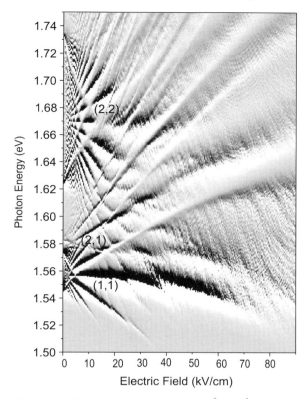

Fig. 7.9. Grayscale map for the 76 Å/39 Å $Al_{0.08}Ga_{0.92}As$ superlattice, showing the calculated joint density of states. The spectra have been differentiated along the energy axis (from [9])

the peak at the upper edge followed by a dip, which can be assigned to the M_1 critical point in the density of states (a saddle point exciton [10]). The miniband width Δ indicated in the figure is (22 ± 2.5) meV.

The spectrum at 9 kV/cm represents the medium-field regime. Different excitonic Wannier–Stark ladder transitions can be clearly resolved. The associated onset of the excitonic ionization continuum for each transition is observed as small steps. The asymmetry in oscillator strength between $(+n)$ and $(-n)$ Wannier–Stark ladder transitions is due to excitonic effects. The hh_{-1} transition gains a considerably higher oscillator strength compared with its counterpart hh_{+1}.

Additionally, a strongly asymmetric line shape of the hh_0 transition, with a slow- and a fast-rising side, is observed. This feature can be attributed to a broadening caused by Fano coupling. Fano resonances occur if a discrete bound state couples to a continuum. The quantum mechanical coupling is

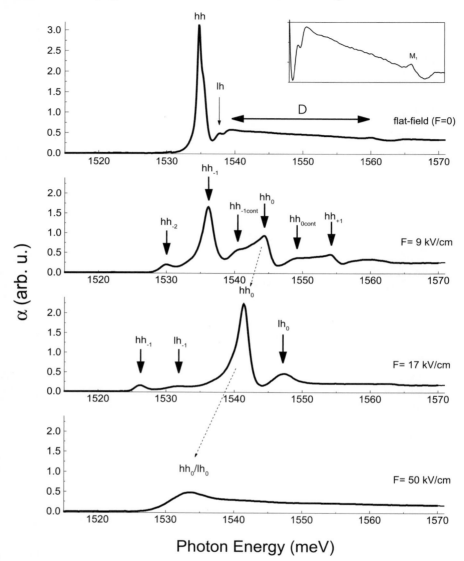

Fig. 7.10. Linear absorption spectra of the 76 Å/39 Å $Al_{0.08}Ga_{0.92}As$ shallow-well superlattice for various electric fields. The *inset* in the flat-field plot shows a close-up of the 1D miniband absorption in the range from 1535 to 1565 meV (from [11])

mediated by the Coulomb interaction. Here, the hh_0 exciton couples to the underlying degenerate continuum of the $hh_{-1}, hh_{-2}, \ldots$ excitons [12].

The field-induced localization of the Wannier–Stark ladder state is demonstrated in the spectrum at 17 kV/cm. The electron wave function is almost localized in one well. The nonvertical Wannier–Stark ladder transitions have nearly vanished and the direct hh_0 transition has gained a high oscillator strength. Also, transitions from light-hole states are clearly resolved.

The localization regime is followed by a field-induced delocalization of the electron wave function. Owing to the interplay of delocalization and tunneling to above-barrier states, a pronounced Zener breakdown is observed. When the field is increased, the hh_0 transition is strongly broadened and has an almost vanishing oscillator strength. This is observed in the high-field spectrum at a field of 50 kV/cm. The lh_0 transition cannot be resolved anymore. For even higher fields (which are not shown here), the absorption vanishes completely.

The linear absorption spectra of the shallow-well sample have been analyzed in order to verify quantitatively the different field regimes discussed in the previous section. The oscillator strength of the hh_0 transition of the (1,1) Wannier–Stark ladder was evaluated as a function of electric field. The oscillator strength is a measure of the electron–hole wave function overlap in the same well, because the hole can be considered to be localized even for high fields [7]. Thus, the oscillator strength monitors the degree of delocalization of the electron wave function.

Figure 7.11 shows the evaluated oscillator strength of the hh_0 absorption transition for the shallow-well sample vs. electric field. The data clearly demonstrate that for low fields, the oscillator strength is low; at medium fields, the field-induced localization of the Wannier–Stark ladder state leads to an increasing oscillator strength of the hh_0 transition, which reaches its maximum at around 25 kV/cm.

This broad maximum has pronounced structure due to anticrossings. At these anticrossings, two degenerate states mix, and the resulting wave function is a superposition of both states. The coupling results in a transfer of oscillator strength from the strong transition to the weak transition in the vicinity of the anticrossing. The oscillator strength has a maximum at the

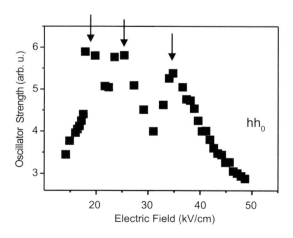

Fig. 7.11. Relative oscillator strength of the hh_0 absorption line vs. electric field. The *arrows* indicate resonant tunneling to above-barrier states as indicated for three transitions by the dashed lines in Fig. 7.8 (from [11])

anticrossing because the transition that has been fitted consists in reality of two or more transitions.

The oscillator strength starts to decrease for fields $F > 35$ kV/cm. This can be attributed to field-induced delocalization of the electron wave function caused by nonresonant tunneling to continuum states. For further increase of the field, i.e. $F > 45$ kV/cm, the oscillator strength of the hh_0 transition almost vanishes (see also the spectra in Fig. 7.10). The field-induced delocalization which is demonstrated here by a decrease in oscillator strength is masked by resonant tunneling to above-barrier states. The field regime investigated here can be described by an interplay of resonant coupling to above-barrier states and a continuous delocalization of the electron wave function. Therefore, the pronounced Zener breakdown is caused by both effects. For high fields, the bound below-barrier state is completely delocalized.

Before concluding the results on linear absorption of superlattices under Zener tunneling conditions, we should remind the reader that superlattices are only quasi-1D systems: excitations in the z direction, which is the growth direction of the superlattice, are coupled to the xy continua. This effect can be removed if the superlattice is held in a large magnetic field, which creates Landau quantization in these continua. Recently, Meinhold et al. [13] have studied Zener tunneling in superlattices in high magnetic fields.

7.1.4 Dynamics of Single Wannier–Stark Ladder States in the Presence of Strong Zener Tunneling

Line Widths In his original paper, Zener [1] treated the tunneling of an electron, described by a Bloch state of one band, into the continuum of Bloch states of a higher-lying band. Zener obtained the following result for the tunneling rate as function of the electric field F:

$$\gamma(F) = \frac{ed|F|}{2\pi\hbar} \exp\left(\frac{-m_e d(\Delta E)^2}{4\hbar^2 e|F|}\right), \qquad (7.2)$$

where d is the superlattice period, m_e the effective electron mass and ΔE the energy gap between the bands. This simple theory predicts that the line widths of the optical transitions in the regime where Zener tunneling sets in should show a smooth increase with electric field; this increase should be exponential for lower fields and linear for higher fields. The increase should be larger when the splitting between the first miniband and the next higher band is smaller.

This prediction has been compared with experimental data taken on two different 35-period shallow superlattices:

- a 76 Å/39 Å GaAs/Al$_{0.08}$Ga$_{0.92}$As sample, having widths of the first and second miniband of 28 and 102 meV, respectively, and a gap ΔE between the first and second miniband of 34 meV (the sample for which results have been described in previous sections); and

- a 50 Å/54 Å GaAs/Al$_{0.11}$Ga$_{0.89}$As sample, having widths of the first and second miniband of 30 and 130 meV, respectively, and a gap ΔE between the first and second miniband of 55 meV.

The key difference between the two samples is the gap between the two minibands; the 50/54 sample has a much larger interminiband gap, leading to weaker coupling and Zener tunneling effects. This is also visible in a theoretical calculation of superlattice absorption based on the theory reviewed by Glück et al. [15]. Figure 7.12 shows the data for the 76/39 sample, which has strong coupling between the minibands. It is obvious that for even low to medium fields, the transitions from the second and higher bands mix with the transitions of the first band, i.e. there is strong Zener coupling.

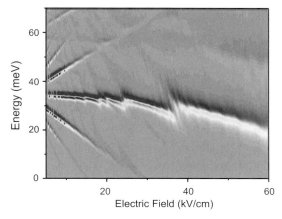

Fig. 7.12. Greyscale map plotting the interband absorption (from the first heavy-hole miniband to the first electron miniband, relative to the bandgap) of the 76/39 sample as function of field, with strong coupling between the first and second electron miniband (from [14])

Figure 7.13 shows the data for the 50/54 sample, with weak coupling between the minibands. Although there is also some coupling between the first miniband and higher bands, the figure shows that this effect is much smaller than for the 76/39 sample, which has a smaller gap between the first and the second band.

We now discuss experimental results on the line width as a function of electric field for the two shallow-well samples. The quantity which was experimentally determined was the width of the vertical $n = 0$ heavy-hole Wannier–Stark ladder transition (hh_0), which was obtained by fitting the absorption line with a Gaussian peak.

Figure 7.14 displays the line width (half width at half maximum, HWHM) of the transition as a function of the internal electric field for the 76/39 sample (with strong coupling). For medium fields, in the Wannier–Stark

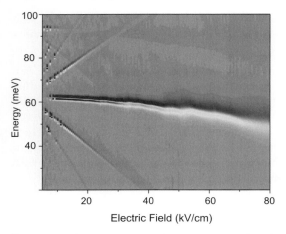

Fig. 7.13. Greyscale map plotting the interband absorption (from the first heavy-hole miniband to the first electron miniband, relative to the bandgap) of the 50/54 sample as function of field, with weak coupling between the first and second electron miniband (from [14])

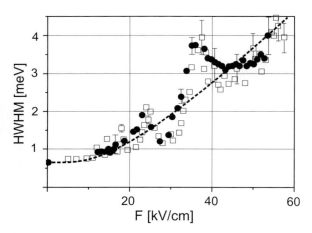

Fig. 7.14. *Solid circles*: line width of hh_0 Wannier–Stark ladder transition of the 76/39 sample, with strong coupling. *Hollow squares*: Fit by theory described in the text [16]. *Solid line*: fit by (7.3) with $A = 0.2$ meV cm/kV, $B = 25$ kV/cm, $C = 0.65$ meV (from [14])

ladder regime, where the electron has a certain probability density in the neighboring wells, the line width stays almost constant. In this field region, the line is Fano-broadened owing to strong coupling to the continuum in the xy direction of lower transitions and can only be fitted approximately.

Here, we concentrate on higher fields at which Wannier–Stark localization of the electronic wave function suppresses the Fano coupling [12] and broadening caused by Zener tunneling starts to become the dominant broadening mechanism [17]. At fields $F > 30$ kV/cm, the line width increases drastically because the coupling to states of higher bands is no longer negligible. In principle, the electron couples to an infinite set of above-barrier Wannier–Stark ladders. Since the energy gap to the first above-barrier band is much larger than the gaps separating the higher above-barrier bands, the coupling can be described in a first approximation by the model put forward by Zener.

We have therefore fitted the overall dependence with the Zener equation (7.2). The finite line width observed at zero field (due to other broadening mechanisms), is approximately accounted for by adding a constant. Using (7.2), the total line broadening is then given by

$$\gamma(F) = C + A|F|\exp\left(-B/|F|\right), \tag{7.3}$$

where $A = d/\pi$, and the constant C accounts for the field-independent line width at zero field. The fit results in parameters A, B with the right order of magnitude, for example $B = 25$ kV/cm, corresponding to $\Delta E \approx 31$ meV.

However, the data in Fig. 7.14 clearly show that sharp oscillations are superimposed on the continuous increase of the line width. These peaks can be attributed to resonant tunneling into states of the first above-barrier miniband; the overall increase of line width is caused by the coupling to all higher bands.

To underline this conclusion, Fig. 7.15 displays the line width as a function of the field for the 50/54, sample with weak coupling. In comparison with Fig. 7.14, the data do not show the pronounced oscillations observed for the 76/39 sample, but mainly a continuous increase of line width, which has been

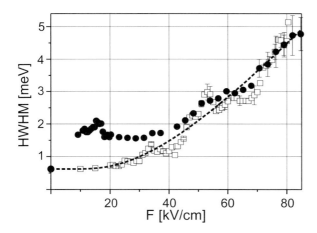

Fig. 7.15. *Solid circles*: line width of the hh_0 Wannier–Stark ladder transition of the 50/54 sample, with weak coupling. *Hollow squares*: Fit by theory described in the text [16]. *Solid line*: fit by (7.3) with $A = 0.21$ meV cm/kV, $B = 60$ kV/cm, $C = 0.62$ meV (from [14])

again fitted by (7.2). Obviously, the small geometric difference between the 76/39 and 50/54 samples results in rather different above-barrier spectra: the gap ΔE between the first and second miniband, which is 20 meV larger, strongly reduces the strength of the tunneling.

The experiments discussed here represent the first direct investigation of the Zener theory in any solid-state system. Surprisingly, there have been previous experiments with atoms in optical lattices (see Chap. 9 for a description of the experimental approach). However, the parameter space in those experiments was smaller than that of the data in Fig. 7.14, so that fewer resonances could be observed [18].

Time-Resolved Experiments The linear absorption results just discussed do not allow one to distinguish between homogeneous and inhomogeneous broadening. Although the linear absorption data presented above suggest that homogeneous broadening dominates at high fields, it is important to quantitatively determine both contributions.

As discussed in Chap. 3, self-diffracted four-wave-mixing experiments are a suitable tool for measuring the dephasing time T_2 if either kind of broadening dominates [19]. Since the sum of the rates of all possible scattering channels is equal to the inverse dephasing time, it should be possible to extract the electron transfer time into above-barrier states if it becomes the dominating mechanism and leads to homogeneous broadening.

Another important issue is to clarify the nature of the broadening caused by Zener tunneling. It is known that the coupling due to the Zener effect removes the discrete energy spectrum of the Wannier–Stark ladder; in the limit of very high fields, the spectrum evolves into a continuum [7]. Therefore, the question arises of which kind of line broadening is observed in the experiment if a strongly Zener-coupled transition (e.g. the hh_0 Wannier–Stark ladder transition) is excited.

Degenerate four-wave-mixing experiments have been performed on the 50/54 shallow superlattice sample (with weak coupling) to measure the dephasing time of the direct hh_0 interband Wannier–Stark ladder transition as a function of the applied electric field. The excitation density was on the order of 10^9 cm^{-2}; the pulse duration was varied from 100 to 300 fs to selectively excite the hh_0 transition. From the observed exponential decay of the four-wave mixing, the dephasing time was determined assuming a homogeneous transition, i.e. that T_2 equals twice the four-wave-mixing decay.

Figure 7.16 shows the dephasing time T_2 as a function of the electric field. For comparison, the inverse line width taken from Fig. 7.15 is also displayed. For low electric fields, crossings with light-hole states and excitonic effects lead to a complex field dependence of the dephasing time. For medium fields, the dephasing time increases owing to the decreasing strength of the Fano coupling [12], which is a broadening mechanism in this field range. For fields $F > 30$ kV/cm, the impact of Zener tunneling as a strong dephasing mechanism seems to be the limiting process for the

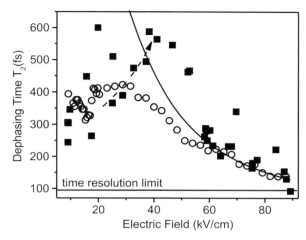

Fig. 7.16. *Squares*: low-temperature (10 K) dephasing time of the hh_0 Wannier–Stark ladder transition as a function of the field of 50/54 sample, with weak coupling. *Open circles*: inverse line width taken from Fig. 7.15. *Dashed arrow*: for medium fields, T_2 increases owing to the reduced Fano effect. *Solid Line*: inverse tunneling rate calculated from (7.2). For high fields, Zener tunneling strongly reduces T_2. The inverse inverse line width agrees well with T_2, indicating homogeneous broadening of the transition (from [14])

lifetime of the Wannier–Stark ladder state. Also, at this field strength the inverse line width corresponds quantitatively to the measured dephasing time. Hence, it can be concluded that Zener tunneling is the dominating dephasing mechanism at these high fields and leads to a purely homogeneously broadened transition. We can conclude that the lifetime of the electronic Wannier–Stark ladder state at high field is directly measured by the four-wave-mixing experiments. An effective transfer time of the electrons into above-barrier states (the Zener tunneling time) can be directly deduced from Fig. 7.16 and is equal to 160 fs, for $F = 75$ kV/cm, for example.

7.2 Wave Packet Dynamics in the Zener Tunneling Regime

7.2.1 Experimental Approach

In the previous section, we discussed the influence of Zener tunneling on a single Wannier–Stark ladder (quasi-)eigenstate. We now address the question of how an electronic wave packet performing Bloch oscillations is influenced by Zener tunneling. At first glance, one might argue that a measurement of this process could only confirm the results of the above section, since the dephasing of the wave packet will be determined by the dephasing of the transitions it is composed of. Surprisingly, the experiments outlined below show that this is not true in all cases.

As shown above (Fig. 7.5), very high fields are needed in deep-well superlattices to achieve strong Zener coupling. As reported in Chaps. 3 and 5, there is no visible influence of Zener tunneling on Bloch oscillations in deep-well superlattices. We therefore discuss results for a shallow-well superlattice again, where we expect that the influence of Zener tunneling on the Bloch oscillations will become significant at lower fields.

For the investigation described below, we selected a shallow-well structure where strong Zener coupling sets in at fields where the nonvertical transitions of the Wannier–Stark ladder still exist. Thus, a wave packet can be excited optically. The sample was the 50 Å/54 Å $Al_{0.11}Ga_{0.89}As$ shallow-well superlattice already discussed above.

As the experimental technique, we chose pump–probe experiments to trace the intraband dephasing time (see Sect. 3.3), i.e. the coherence between the electronic states that the Bloch wave packet is composed of. To obtain a high signal-to-noise ratio, differential pump–probe signals were measured by using a shaker which wiggled the temporal delay, and a lock-in amplifier for detection.

For a qualitative comparison, Fig. 7.17 displays both pump–probe data and four-wave-mixing data for the 50/54 sample. It shows data gathered at an internal field of 11 kV/cm on the shallow-well superlattice. As expected, the pump–probe derivative signal crosses the amplitude axis at zero delay. The pump–probe signal can easily be observed for twice as long as the four-wave-mixing signal. In fact, coherence was observed in this pump–probe trace up to 3.5 ps, while in four-wave mixing a coherent signal was obtained only up to 1.3 ps.

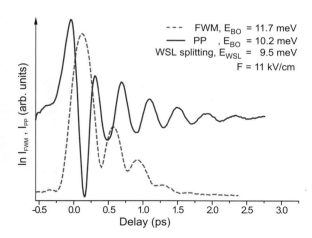

Fig. 7.17. Pump–probe and four-wave-mixing signals from the 50/54 shallow-well sample. The four-wave-mixing plot was derived by integrating spectrally over hh_{-1} (from [20])

The pump–probe signal shown in Fig. 7.17 decays with a time constant

$$\tau_{\text{intra}} = \frac{1}{\gamma_{\text{intra}}} = 700 \text{ fs}, \tag{7.4}$$

determined by the intraband coherence decay rate γ_{intra} while the four-wave-mixing signal decays with a time constant

$$\tau_{\text{inter}} = \frac{1}{2\gamma_{\text{inter}}} = 505 \text{ fs}, \tag{7.5}$$

determined by the interband coherence decay rate γ_{inter} between the Wannier–Stark ladder states in the conduction and valence bands.

These decay times were determined by fitting an exponentially decaying harmonic oscillation to the signals. Fitting the decay of the modulation of the pump–probe trace is easily accomplished, while the four-wave-mixing signal does not decay linearly in its semilogarithmic representation, and thus an error on the order of 100 fs has to be assigned to τ_{FWM}. Of course, the first oscillation period around zero delay has been excluded from the fits for both transients to avoid the temporal interval of pulse overlap, where coherently scattered light enhances the signal. To gain a sufficient signal-to-noise ratio, an excitation density (for both beams together) of $\sim 2 \times 10^9 \text{cm}^{-2}$/well was used.

7.2.2 Field-Induced Decay of the Intraband Polarization

To investigate the influence of Zener tunneling on the dynamics of a Bloch-oscillating wave packet, the damping of the pump–probe signal was measured over a large field range. As discussed above, the damping of the pump–probe data directly reveals the intraband decay time.

Figure 7.18 shows the pump–probe transients. Signal modulation can be observed up to 4 ps at low applied fields. The internal field was derived from the spectral position of the hh_{-1} peak in the simultaneously obtained spectrally resolved four-wave-mixing signal at zero delay.

Figure 7.19 shows the tunability of the Bloch oscillation period with increasing electric field. The inverse period scales approximately linearly with the field, as expected, and demonstrates once more the surprisingly large tunability range of the Bloch oscillator. Note that the pump–probe beating effect does not fully agree with the splitting of the Wannier–Stark ladder. However, those splittings are not equal for all transitions in any case, owing to excitonic effects.

Figure 7.20 shows the dependence of the intraband damping time $\tau_{\text{intra}} = \frac{1}{\gamma_{\text{intra}}}$ on the electric field. The intraband damping time describes the coherence time between the two simultaneously excited Wannier–Stark ladder states hh_0 and hh_{-1}. The decay constants were obtained from the pump–probe data presented in Fig. 7.18.

142   7 Damping of Bloch Oscillations II: Zener Tunneling

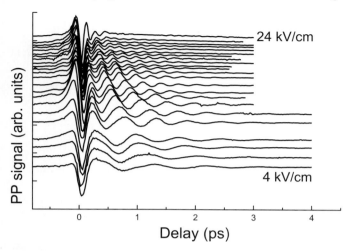

Fig. 7.18. Pump–probe transients for an internal field tuned from 4 kV/cm to 24 kV/cm. The laser was centered between hh_{-1} and hh_0. The excitation density was 2×10^9 cm^{-2} per well with collinear polarization. The lattice temperature was 8 K (from [20])

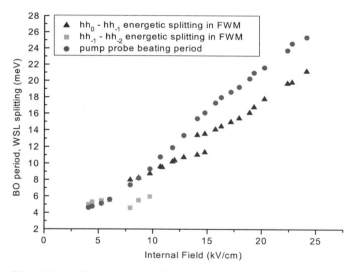

Fig. 7.19. Frequency of Bloch oscillations derived from the pump–probe data, compared with the splittings of the Wannier–Stark ladder in the spectrally resolved four-wave-mixing data at zero delay (from [20])

A continuous decrease in the intraband decay time with increasing field is seen. This is followed by an abrupt decrease of the intraband decay time, which sets in at about 24 kV/cm. To proceed further, we shall consider a 3D

7.2 Wave Packet Dynamics in the Zener Tunneling Regime

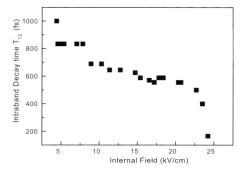

Fig. 7.20. Decay constant of the exponential decay of the intraband polarization. The data were derived from the modulation amplitude of the pump–probe transients shown in Fig. 7.18 (from [20])

plot of the four-wave-mixing spectra vs. the electric field. This chart is shown in Fig. 7.21 and provides the spectral details not resolvable in linear absorption which allow one to interpret the behavior of the intraband coherence. The evolution of the transitions hh_{-2}, hh_{-1}, and hh_0 with increasing field F can be discerned. At ≈ 19 kV/cm, an avoided crossing of lh_{-1} with hh_0 is indicated by a longer arrow. The avoided crossing and the corresponding energetic splitting of about 2.5 meV can be seen in the four-wave-mixing

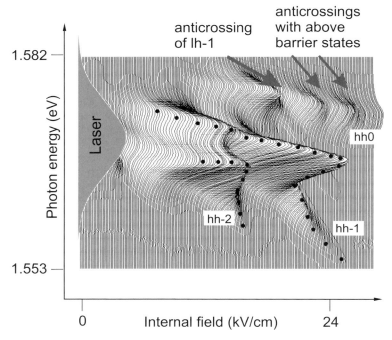

Fig. 7.21. Four-wave-mixing spectra as a function of electric field for zero delay time and constant spectral position of the laser (from [20])

spectra. Obviously, the anticrossing of two below-barrier states does not have a pronounced influence on the Bloch wave packet dynamics, since there is no change of the Bloch damping at this field.

The situation is different for the first anticrossing of above-barrier states, which becomes visible at ≈ 24 kV/cm (two short arrows). The crossings from above-barrier states with hh_0 were predicted by theory and thus also show up in the calculated fan chart of the Wannier–Stark ladder transitions (not shown). This crossing causes a dramatic reduction of the Bloch-oscillation damping time. Thus, the data in Fig. 7.20 show the first observation of *the damping of Bloch oscillations due to Zener breakdown*. This effect is masked in deep superlattices by interface and phonon scattering [21].

The experimental results were compared with a direct numerical solution of the semiconductor Bloch equations (SBEs). In particular, in order to properly describe Zener tunneling processes in the high-field regime, the generalized SBEs formulation described in [22, 23] (which includes intra- as well as inter-miniband electric-field contributions) was employed.

Figure 7.22 shows typical theoretically calculated oscillation traces below (20 kV/cm) and above (33 kV/cm) the threshold field of the experiment. Below this field, Bloch oscillations are basically undamped in the theoretical model. For fields larger than the threshold, a very fast decay due to resonant Zener tunneling, with a decay time of about 200 fs, is predicted. This result is quite close to the experimentally obtained value of about 160 fs.

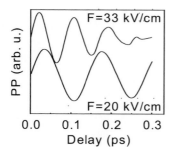

Fig. 7.22. Theoretical results for Bloch oscillation dynamics. The traces show the results for fields below (20 kV/cm) and above (33 kV/cm) the threshold field (from [11])

The theoretical decay times exhibit a step behavior similar to that observed in the experiment, but with a somewhat larger threshold field. This discrepancy can be ascribed to a number of phenomena not included in our theoretical modeling, for example intra- and interband Coulomb effects, and band nonparabolicity. Also, one has to take into account the fact that the precision of the experimental determination of the field is limited.

It is quite interesting to compare the interband dephasing time of a *single* Wannier–Stark ladder state with the intraband dephasing of the Bloch-oscillating wave packet discussed here: if one compares the data of Fig. 7.16 with the intraband data of Fig. 7.20, one can recognize that the *intraband decay time falls below the interband decay time.* This is a quite surprising result, since one would expect at first sight that the dephasing time of the wave packet would be given by the dephasing time of its constituents, the Wannier–Stark ladder states. If one measures the interband coherence of those states, one will measure a sum of the scattering processes of the electrons and the holes. A measurement of the intraband dephasing time of the wave packets should be influenced only by the electron scattering processes and should thus typically be longer than the interband time (and only be identical to the interband time if hole scattering is negligible). Here, however, the intraband time is *shorter* than the interband time, which cannot be understood in this simple single-particle picture.

One possible speculation to explain this is that the wave packet in the intraband experiments performs a semiclassical trajectory which leads to very high scattering probabilities if, for example, it has reached its maximum field-downward direction. This might be compared to the situation for Rydberg wave packets in atoms, where the field ionization probability varies periodically with time, a variation which is used to follow the spatial motion of the wave packet [24].

7.2.3 Wave Packet Revival in the Presence of Zener Tunneling

Up to this point, we have used a simple picture in which Zener breakdown is due to resonant tunneling from hh_0 to the above barrier states. This picture is particularly suggestive, because the energetic degeneracy of hh_0 and the above-barrier states is seen in the fan chart. However, if one considers the effect of Zener tunneling on Bloch oscillations, one has to deal with a whole wave packet of Wannier–Stark ladder transitions. In the following, we show that the resonant Zener coupling can lead to a complicated spatial revival of the Bloch wave packet.

We discuss here the experimental observation and numerical simulation of a polarization revival due to spatial wave packet decay and re-formation in a shallow superlattice [25]. The physical origin of this wave packet revival in a solid is somewhat different from that of the Rydberg wave packet revivals reported in atomic physics [26], where the wave packet dephasing is, in general, due to both energy dispersion of the nearly equally spaced Rydberg states and the different anharmonic effective potentials which the respective Rydberg states are subject to. The wave packet, made of high-quantum-number states, spreads from a classically localized particle to become nonclassical and, after the revival time, rephases and travels on its classical trajectory again. Here, owing to the discrete number of states excited, the loss of phase caused by the anharmonicity of the atomic potential can be fully recovered; after several oscillatory cycles (with a period determined by the energy spacing between the Rydberg states), a full revival can occur. The revival time is proportional to the number of excited states.

In the superlattice case discussed here, a wave packet made of localized Wannier–Stark ladder states from the first miniband and nearly degenerate, delocalized above-barrier states from the second miniband is optically excited. The revival originates from a quantum beating between these few discrete states. Figure 7.23 shows, in the left panel, selected high-field traces from Fig. 7.18. On the trace for 24 kV/cm, a recommencing of oscillations is visible at 1.6 ps. This recommencing is interpreted as a revival of the macroscopic intraband polarization. The exact revival time is difficult to estimate, because neither the exact zero delay nor the actual amplitude maximum of the revival can be specified precisely.

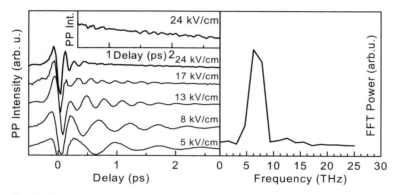

Fig. 7.23. *Left*: selected traces from Fig. 7.18 for the high-field range. The uppermost transient shows a recommencing of the oscillations after a range of destructive interference (shown enlarged in the inset). *Right*: Fourier transform of the revival. The temporal range from 1.5 to 2 ps was analyzed (from [20])

The Fourier transform in the right panel of Fig. 7.23 clearly demonstrates the quantum beat origin of the revival. The maximum at 6.25 THz in frequency space, equivalent to a beat period of 160 fs, equals the Bloch-oscillation beating period of 163 fs within the first 500 fs of delay, as obtained earlier from the decay fits. It is thus obvious that the weak modulation after about 1.6 ps is again due to Bloch oscillations in the first miniband. The oscillations in the second miniband are not visible in the pump–probe traces since there is no optical coupling to the holes generated in the first miniband by the optical excitation.

Figure 7.24 shows a schematic description of the experiment. The wave packet is first excited in the first miniband and performs Bloch oscillations with a time period given by the inverse of the Wannier–Stark splitting. However, owing to the anticrossing with the states of the second miniband, the wave packet also contains wave functions of the second miniband. With a time constant given by the inverse of the much smaller splitting due to this anticrossing of minibands 1 and 2, the wave packet moves to the second band

7.2 Wave Packet Dynamics in the Zener Tunneling Regime   147

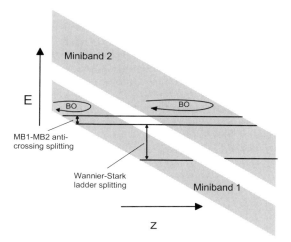

Fig. 7.24. Schematic illustration of the spatial dynamics when a wave packet revival occurs. The wave packet first performs Bloch oscillations in the first miniband; it then transfers to the second miniband (owing to the superposition of the anticrossed states) to perform Bloch oscillations there, and finally returns to the first miniband

and performs Bloch oscillations there with a much larger amplitude, but still with the same frequency. For the shallow superlattice samples investigated here, the Bloch oscillations in the second band occur between unconfined states above the barriers. This sequence repeats until the carriers have lost their coherence.

To model these experimental results quantitatively, the confined and the above-barrier electronic states of a 35 well superlattice, subject to a static electric field, have been calculated [25]. A plane wave expansion method [27, 28] in the framework of effective-mass envelope theory was used. The plane waves of the ideal bulk material form an orthonormal, complete basis set, and thus can be used to expand the envelope function of a biased superlattice. Implicitly, the use of the plane wave basis for a finite superlattice means that periodic boundary conditions are adopted. The localized and delocalized states can be obtained by directly diagonalizing the Hamiltonian matrix.

The simulation shows that at a threshold electric field, the confined Wannier–Stark ladder-states nearly overlap energetically and, to some extent also overlap spatially with delocalized states that lie above the barrier at flat field (shown in Fig. 7.25). This resonance was directly observed experimentally as an avoided crossing of the hh_0 transition with a transition to an above-barrier state in spectrally resolved four-wave mixing (Fig. 7.21).

We now discuss the wave packet motion associated with the optical excitation of the states discussed above. At 24 kV/cm, the Bloch-oscillating wave packet is formed by two below-barrier states of the first miniband. Such a wave packet is created simultaneously in each well by optical excitation. To visualize the spatial behavior, we discuss one specific packet, centered on wells 12 and 13. The calculated eigenenergies are 345.3 meV and 370.3 meV with respect to the band bottom. Owing to the anticrossing with the higher transitions, each confined state is accompanied by an above-barrier state

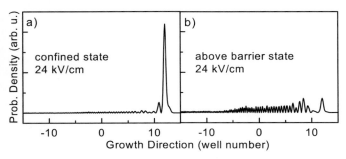

Fig. 7.25. Calculated electron spatial probability density of (**a**) a confined state of the 12th superlattice well and (**b**) the associated above-barrier state at 24 kV/cm. The two states partially overlap spatially and have an energetic splitting of about 1.65 meV, which gives a theoretical revival period of about 2.5 ps (from [25])

shifted by 1.65 meV in energy from the confined state. The above-barrier state associated with the confined state of well 12 is dispersed largely over a range from well 7 to well 10 but has its density spread over more than 12 lattice periods. The above-barrier state associated with the confined state of well 13 is shifted by one lattice period, relative to the state associated with well 12.

As described above, the optical excitation of these four states then creates a wave packet which has two built-in periods: a short period corresponding to the splitting of the Wannier–Stark ladder states (i.e. the Bloch frequency), and a long period corresponding to the small splitting of 1.65 meV, which is due to the anticrossing of the first- and second-miniband states.

The wave packet then repeatedly Zener-tunnels resonantly between discrete confined states and discrete above-barrier states within the theoretical revival time of 2.5 ps (the experimental value is 1.8 ps). The wave packet performs Bloch oscillations in both the first and the second miniband, with the same frequency $\omega_\mathrm{B} = eFd/\hbar$. However, the spatial amplitude $L = \Delta/eF$ is much larger in the second band, owing to the larger miniband width.

If one visualizes the motion of the electron using the theoretical model, it becomes obvious that the electron performs a quite interesting spatial motion: the electron probability is sequentially transferred spatially over more than 10 wells, i.e. about one hundred nanometers. Figure 7.26 visualizes the spatial dynamics of this wave packet.

From our results, it follows that the observed initial drop in the wave packet coherence time is due to destructive quantum interference, caused by the confined states and above-barrier states. Besides the Bloch beating within one band, a beating of confined states with delocalized above-barrier states is observed. In other words, the Zener tunneling retains coherence when the electron tunnels to one discrete state only. It would be interesting to extend these studies to higher fields where the Bloch wave packet couples to

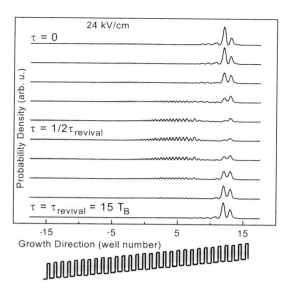

Fig. 7.26. Calculated time evolution of the spatial probability density of a Bloch-oscillating wave packet composed of two confined and two above-barrier states, from $\tau = 0$ to $\tau = 2.5$ ps (from [25])

several higher transitions, leading to a polarization decay due to decoherence, without rephasing on a reasonable timescale.

7.3 Summary

In this section, we have discussed dephasing by coupling to higher bands (Zener tunneling). In the early years of studies on Bloch oscillations, this problem was largely neglected, although the proposal of Bloch oscillations was originally made in conjunction with Zener tunneling.

It turns out that in standard semiconductor superlattices, Zener tunneling is a comparatively weak effect and does not significantly contribute to the damping of Bloch oscillations. In shallow-well superlattices, however, Zener tunneling is a significant damping mechanism and can be studied in a detailed manner, including a determination of the dynamics. The results are in excellent agreement with theoretical calculations of the electronic structure in the presence of a bias field.

References

1. C. Zener, Proc. R. Soc. London Ser. A **145**, 523 (1934).
2. L. Esaki, Phys. Rev. **109**, 603 (1958).
3. H. Schneider, H.T. Grahn, K. von Klitzing, and K. Ploog, Phys. Rev. Lett. **65**, 2720 (1990).
4. A. Sibille, J.F. Palmier, and F. Laruelle, Phys. Rev. Lett. **80**, 4506 (1998).

5. M. Helm, W. Hilber, G. Strasser, R. De Meester, F.M. Peeters, and A. Wacker, Phys. Rev. Lett. **82**, 3120 (1999).
6. S. Glutsch and F. Bechstedt, Phys. Rev. B **57**, 11887 (1998).
7. S. Glutsch and F. Bechstedt, Phys. Rev. B **60**, 16564 (1999).
8. J.E. Avron, J. Zak, A. Grossmann, and L. Gunther, J. Math. Phys. **18**, 918 (1977).
9. B. Rosam, Diplomarbeit, Technische Universität Dresden (1999).
10. N. Linder, Phys. Rev. B **55**, 13664 (1997).
11. B. Rosam, D. Meinhold, F. Löser, V.G. Lyssenko, S. Glutsch, F. Bechstedt, F. Rossi, K. Köhler, and K. Leo, Phys. Rev. Lett. **86**, 1307 (2001).
12. C.P. Holfeld, F. Löser, M. Sudzius, K. Leo, D.M. Whittaker, and K. Köhler, Phys. Rev. Lett. **80**, 874 (1998).
13. D. Meinhold, K. Leo, N.A. Fromer, D.S. Chemla, S. Glutsch, F. Bechstedt, and K. Köhler, Phys. Rev. B **65**, 161307 (2002).
14. B. Rosam, D. Meinhold, M. Glück, H.J. Korsch, K. Leo, and K. Köhler, submitted to Phys. Rev. B, 2003.
15. M. Glück, A.R. Kolovsky, and H.J. Korsch, Phys. Rep. **366**, 103 (2002).
16. M. Glück, A.R. Kolovsky, H.J. Korsch, and F. Zimmer, Phys. Rev. B **65**, 115302 (2002).
17. S. Glutsch, F. Bechstedt, B. Rosam, and K. Leo, Phys. Rev. B **63**, 085307 (2001).
18. S.R. Wilkinson, C.F. Bharucha, K.W. Madison, Q. Niu, and M.G. Raizen, Phys. Rev. Lett. **76**, 4512 (1996).
19. T. Yajima and Y. Taira, J. Phys. Soc. Jpn. **47**, 1620 (1979).
20. D. Meinhold, Diplomarbeit, Technische Universität Dresden (2000).
21. F. Rossi, M. Gulia, P.E. Selbmann, E. Molinari, T. Meier, P. Thomas, and S.W. Koch, Proc. 23rd Int. Conf. Phys. Semicond., Berlin, 1996, eds. M. Scheffler and R. Zimmermann (World Scientific, Singapore 1996), p. 1775.
22. F. Rossi, Semicond. Sci. Technol. **13**, 147 (1998).
23. F. Rossi, "Bloch Oscillations and Wannier–Stark Localization in Semiconductor Superlattices", in *Theory of Transport Properties of Semiconductor Nanostructures*, ed. E.E. Schöll (Chapman and Hall, London 1998).
24. M.L. Biermann and C.R. Stroud, Jr., Phys. Rev. B **47**, 3718 (1993).
25. D. Meinhold, B. Rosam, F. Löser, V. G. Lyssenko, F. Rossi, J.-Z. Zhang, K. Köhler, and K. Leo, Phys. Rev. B **65**, 113302 (2002).
26. J. Yeazell and T. Uzer, eds., *The Physics and Chemistry of Wave Packets* (Wiley, New York 1999).
27. B.F. Zhu and Y.C. Chang, Phys. Rev. B **50**, 11932 (1994).
28. J.Z. Zhang, B.F. Zhu, and K. Huang, Phys. Rev. B **59**, 13184 (1999).

8 Electrical Transport Experiments

In the previous Chapters of this book, we have discussed experiments on Bloch oscillations and Zener tunneling which were performed by optical excitation of the electronic wave packets. Despite the elegance of this approach, one has to realize that the "real beef" in semiconductors is concepts which will work in simple devices. This is obviously not the case for approaches using ultrafast optical excitation.

We therefore address here the electrical transport properties of semiconductor superlattices. The first transport experiments were performed soon after the first superlattice structures were grown.

First, we discuss in Sect. 8.1 the concept of miniband transport and the various theoretical models which can be applied to describe it. These models are compared with experimental results on general aspects of miniband transport (Sects. 8.2 and 8.3). Chapter 8.4 covers some more detailed aspects of miniband transport. Then, we discuss experiments on Bloch oscillations in SiC, which is a natural superlattice (Sect. 8.5).

In Sect. 8.6, we discuss experiments where transport studies were combined with irradiation of the samples with electromagnetic radiation. This section also discusses experiments which investigate Zener tunneling by transport techniques.

8.1 Modeling of Transport in Superlattices

A number of more or less sophisticated theoretical models have been put forward to describe miniband transport. A lucid discussion of these models has been given by Sibille [1]. We give here a simplified description along those lines. We concentrate first on band transport models. As we shall discuss later (Sect. 8.6.1), there are also other theoretical approaches to explain the negative differential velocities, for example hopping and sequential tunneling.

8.1.1 Semiclassical Miniband Transport Theories

Esaki–Tsu Model The simplest transport model was put forward by the inventors of the semiconductor superlattice, Esaki and Tsu [2]. The theory starts out with the acceleration theorem

$$\hbar \frac{dk}{dt} = eF ,\tag{8.1}$$

which predicts spatially harmonic Bloch oscillations as discussed in Chap. 1. In particular, in the case of a harmonic energy dispersion of the miniband

$$E(k) = \frac{\Delta}{2} - \frac{\Delta}{2}\cos(kd) ,\tag{8.2}$$

the group velocity v is given by

$$v_{\mathrm{R}} = \frac{1}{\hbar}\frac{\partial E}{\partial k} = \frac{d\Delta}{2\hbar}\sin(kd) .\tag{8.3}$$

The following approach, which we term the Esaki–Tsu model, introduced scattering processes: Instead of being transported ballistically for an infinite time, a carrier will scatter after a finite time τ. Note that this scattering time is independent of the energy of the carrier in this simple model.

The average velocity of the carrier ensemble is then obtained by the integration

$$\begin{aligned} v &= \int_0^{+\infty} \exp(-t/\tau)\, dv_{\mathrm{R}} \\ &= \frac{eF}{\hbar^2}\int_0^{+\infty} \frac{\partial^2 E}{\partial k^2}\exp(-t/\tau) dt ,\end{aligned}\tag{8.4}$$

which with the assumption of a harmonic dispersion (8.2), yields

$$v = \frac{\mu F}{1 + (F/F_{\mathrm{C}})^2} ,\tag{8.5}$$

where

$$\mu = \frac{e\Delta\tau d^2}{2\hbar^2} \quad \text{and} \quad F_{\mathrm{C}} = \frac{\hbar}{e\tau d}\tag{8.6}$$

are the mobility and the critical field, respectively.

In the low-field case $F \ll F_{\mathrm{C}}$, the velocity is proportional to the field, and the constant of proportionality is the mobility μ. If one assumes the effective mass corresponding to a harmonic dispersion (8.2), one obtains the standard Drude result

$$\mu = \frac{e\tau}{m^*} ,\tag{8.7}$$

as discussed in Chap. 1.

The derivative of (8.5) with respect to F immediately yields the result that there is a peak velocity $v_{\max} = \Delta d/(4\hbar)$ at $F = F_{\mathrm{C}}$; after the peak, the velocity starts to decrease with field, i.e. negative differential velocity occurs. The reason for the behavior can be understood if one calculates the ensemble average of the $\boldsymbol{k}$-vector, which turns out to be

$$\langle k \rangle = \frac{eF\tau}{\hbar}. \tag{8.8}$$

At the critical field F_C, we obtain $\langle k \rangle = 1/d$. The average $\boldsymbol{k}$-vector is still well below the inflection point of the band at $\boldsymbol{k} = \pi/2d$, but the contribution of the part of the ensemble having negative effective mass is large enough to cause a decrease in drift velocity for a further increase of the electric field.

Approximate Solutions of the Boltzmann Equation Although it contains the key features of miniband transport, the Esaki–Tsu model contains strong simplifications and does not allow quantitative conclusions. To achieve such conclusions, the following simplifications must be removed:

- The Esaki–Tsu model is one-dimensional, i.e. it neglects the fact that carriers can gain $\boldsymbol{k}_x$ and $\boldsymbol{k}_y$ momentum owing to scattering.
- In the Esaki–Tsu model, collisions always put the electron back to $\boldsymbol{k} = 0$.
- The thermal broadening of the carrier ensemble is not taken into account.
- The model describes the various scattering mechanisms in a solid by a single, constant scattering time τ.

In principle, all these shortcomings can be removed by using the Boltzmann equation. However, the inclusion of all effects leads to equations which have no analytical solution [1]. The simplest approach is to use the Boltzmann equation in the relaxation time approximation, where the distribution function f obeys

$$\frac{\partial f}{\partial t} + \frac{eF}{\hbar}\frac{\partial f}{\partial k_z} = f_0/\tau, \tag{8.9}$$

where f_0 is the equilibrium thermal distribution. With a thermal distribution in the x and y directions, the Boltzmann equation can be solved for the z direction, and finally yields the following for the ensemble average velocity [3]:

$$v = \frac{I_1(\Delta/2k_\mathrm{B}T)}{I_0(\Delta/2k_\mathrm{B}T)} \frac{\mu F}{1 + (F/F_\mathrm{C})^2}, \tag{8.10}$$

where I_1 and I_0 denote modified Bessel functions.

Compared with the Esaki–Tsu theory, the only generalization in the relaxation time approximation is that the thermal broadening is included. It is obvious that the only modification in the equation, compared with the Esaki–Tsu theory, is the appearance of a Bessel function expression, which is multiplied by the velocity (8.5). The Bessel function term approaches 1 when $k_\mathrm{B}T$ is much smaller than the miniband width Δ, i.e. in the low-temperature limit. For high T, it scales as $1/T$, i.e. the ensemble average velocity goes to zero when the thermal energy is much larger than the miniband width. The origin of this behavior is the fact that at high temperature, the electrons are equally distributed over the band, and there is no significant effect of the field on the distribution. In this case carriers with both positive and negative mass are present in equal numbers, and the ensemble drift velocity tends towards zero. As we shall discuss below, this thermal saturation has been observed by Brozak et al. [4,5].

Full Solutions of the Boltzmann Equation If one wants to avoid the other simplifications of the Esaki–Tsu model, one needs to obtain a solution of the full Boltzmann equation. This basically means that one has to replace the relaxation term f_0/τ by a full scattering term which includes the various scattering processes. The resulting equations are obviously complicated. This form of the Boltzmann equation has been solved using Monte Carlo techniques, where the ensemble average is determined by following the statistical path of a limited number of carriers.

Figure 8.1 shows the result of such a Monte Carlo calculation [6]. It is obvious that the general behavior is quite similar to the predictions of the Esaki–Tsu model. A quantitative comparison (see [1]) shows that the Esaki–Tsu model agrees reasonably with the full Boltzmann model when the critical field, for example, is calculated.

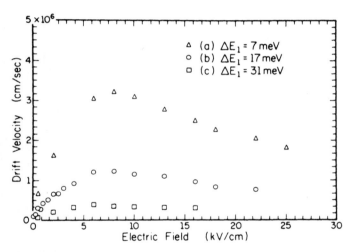

Fig. 8.1. Electron drift velocities in three different AlGaAs/GaAs superlattices with miniband widths of 7, 17, and 31 meV as a function of electric field. Note that the labeling is erroneous: the highest velocity is reached for the widest miniband, with 31 meV band width (from [6])

8.1.2 Scattering Mechanisms in Superlattices

The two scattering mechanisms which determine the mobility in bulk semiconductors are ionized-impurity scattering and phonon scattering. Both processes are also present in superlattices, although the latter might be modified

8.1 Modeling of Transport in Superlattices   155

by changes in the phonon system because of the superlattice structure. Additionally, the unavoidable interface roughness is another factor which contributes as a scattering process to the transport properties of superlattices. Interface roughness has been addressed by some authors, with simplifying assumptions about the nature of the roughness. The simplest model was by Chomette and Palmier [7], who have calculated the scattering rates due to a local fluctuation of the barrier/well thickness of one monolayer. The scattering rate was calculated as a function of the radius of such a thickness fluctuation defect; the defect density was kept constant (see Fig. 8.2). The data display a minimum in the mobility, i.e. a maximum of the scattering rate, at a defect radius of about 100 nm.

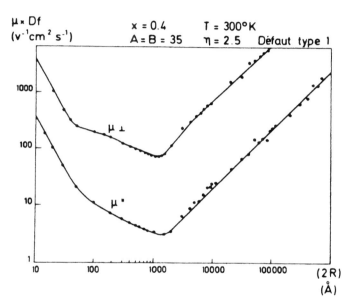

Fig. 8.2. Longitudinal and transverse mobility (as determined by interface roughness scattering) in a 35 Å/35 Å AlGaAs/GaAs superlattice ($x = 0.4$) as a function of the roughness diameter (from [7])

An improved model of interface roughness, by Dharssi and Butcher, considered interface roughness with a Gaussian correlation spectrum of the scattering matrix element between miniband Bloch functions, parametrized by the coherence length of the interface roughness [8]. Figure 8.3 displays the drift velocity vs. field for superlattice structures with a given miniband width, for various relations between well width (parameter a) and barrier width (parameter b), given in Å. It is immediately obvious that the scattering for a structure with a wide well is weak, leading to a pronounced maximum at the critical field. For a narrow well and a broad barrier, the velocity curve

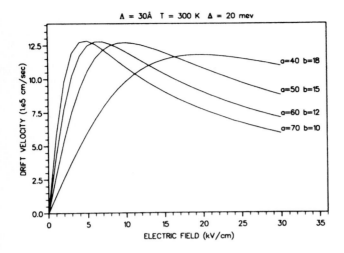

Fig. 8.3. Drift velocity as a function of field for different superlattices (from [8])

is smeared out, indicating that the high scattering rate prevents negative differential velocity and Bloch oscillations. This result strongly suggests that one should use wide-well superlattices, which is confirmed by the results in Chaps. 3 and 4. The model of Dharssi and Butcher was used by Palmier et al. for calculations based on the Boltzmann equation [9].

8.2 Experimental Investigations of Miniband Transport

8.2.1 Investigations of Low-Field Transport

Optical Techniques In Chaps. 3 to 5, we discussed how coherent optical techniques can be used to trace Bloch oscillations in superlattices. It is also possible to study ensemble drift transport using optical techniques. In these experiments, the basic idea is that the carriers can be generated at a rather well-defined location and that their transport can be monitored by some change in the optical response.

One example of the power of such techniques is provided by the experiments done by Deveaud and coworkers, which used photoluminescence of superlattices to access perpendicular electron and hole transport by the use of specially designed structures [10–13]. The basic idea was to include an enlarged well in the superlattice, which serves as an efficient trap for the photoexcited carriers. The arrival of carriers at the enlarged well was then detected by photoluminescence. The enlarged-well technique can be combined with a graded-gap superlattice, where the band gap changes from well to well, thus causing drift transport without an external applied field. Note that owing to the detection by photoluminescence, incoherent carrier ensembles can be detected, in contrast to the four-wave-mixing and THz techniques discussed previously.

Despite their simplicity and elegance, these optical methods have some drawbacks which need to be considered:

- As we have learned in the previous chapters on experiments with optical excitation, a considerable modification of the electric field by the carrier generation is possible.
- In optical experiments, electrons and holes are generated. The system then contains both types of free carriers and additionally excitons, and it is not evident a priori which kind of transport dominates. By investigating doped samples, it is possible to circumvent this problem [13].
- It is well known that the trapping of carriers in potential wells might have rather slow dynamics, depending on various factors such as the type of carrier [14,15]. This introduces a number of parameters into the determination of the dynamics.

The simplest approach in these optical transport experiments uses continuous-wave photoluminescence. In this case, transport parameters can be derived by comparing the intensity from the superlattice and the photoluminescence of the enlarged well. However, such an analysis is rather indirect. Much more direct information can be obtained by time-resolved experiments.

As an example, we discuss here a time-resolved experiment by Deveaud et al. [10]. In this experiment, a graded-gap superlattice and an enlarged well were combined. The sample structure is shown in the inset of Fig. 8.4 and consists of ten graded-gap wells and an enlarged well. Despite the broadening, some individual wells can be discerned in the spectra, thus allowing one to follow the ambipolar transport across the superlattice structure. For short delay times, the spectra are dominated by the signal from the first wells, close to the surface; for later delay times, the center shifts to the enlarged well, where most of the carriers have arrived after about 100 ps. The spectra after that delay time resemble in the overall shape the continuous-wave spectra.

Deveaud et al. [10] have derived the ambipolar mobilities from these data. They obtained mobilities up to $\mu = 1800$ cm^2 / V s, which clearly shows that the perpendicular transport is Bloch-like transport in the miniband and not hopping transport from well to well. Recently, these measurements have been extended [16], on the basis of a graded-gap structure including codoping to lower the gap in a particular well [17].

Electrical Techniques Here, we briefly discuss experiments which employ transistor structures to measure the mobility in a superlattice in the low-field regime. The first such experiment was performed by Palmier et al. [18]. The idea of this experiment was as follows. The superlattice structure was inserted into the base of an npn heterobipolar transistor. Owing to the efficient carrier injection from the emitter, the gain of such a transistor is mainly

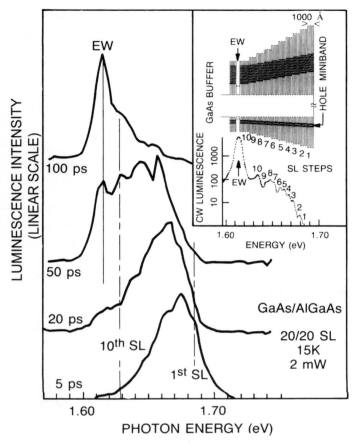

Fig. 8.4. Time-resolved photoluminescence spectra in a graded-gap superlattice structure with ten wells and an enlarged well. *Inset*: sample structure and continuous-wave photoluminescence spectrum. In the 20/20 sample shown here, the Al content in the barriers of the 20/20 Å $GaAs/Al_xGa_{1-x}As$ sample varies from $x = 0.35$ to $x = 0.17$ (from [10])

determined by the transport through the base. If the minority-carrier lifetime is well known, one can determine the diffusion length and, from the Einstein relation, the mobility of the electrons in the base. Because the recombination times were not known well, the analysis relied on a numerical model of the transistor action. With reasonable assumptions on the lifetime, an electron superlattice mobility of about 1000 cm^2/(Vs) was found in a 4.5 nm/2.0 nm $GaAs/Al_{0.27}Ga_{0.73}As$ structure, and a mobility of less than 10 cm^2/(Vs) for a similar structure with a 4.0 nm wide barriers. This marked difference for otherwise identical structures supports the assumption that the superlattice transport was the most relevant parameter in the structures investigated.

8.2 Experimental Investigations of Miniband Transport

The very low mobility for the second structure indicates that for this weakly coupled structure, band transport is no longer relevant since the disorder leads to hopping transport.

8.2.2 Resonances with Higher Minibands: Domain Formation

In their first experiments investigating negative differential velocity in superlattices, Esaki and Chang [19] observed a series of transport resonances in a 4.5 nm well/4.0 nm barrier GaAs/AlAs superlattice with narrow bands, with a voltage period which corresponded roughly to the minigap between the first and second electron miniband. The interpretation of this effect is the formation of electric-field domains in the sample: there is a high-field domain where the first miniband is in resonance with the second miniband of the adjacent well, and a low-field domain where the states of the first miniband are in resonance. A similar situation can exist for a resonance of the first miniband with the third or a higher miniband. Thus, it is possible that several different domains are simultaneously present in the sample.

When the voltage is increased, the position of the domain boundary moves to accommodate the additional potential drop: for a certain voltage step, the domain boundary moves by one well so that the high-field domain increases in width by one well. The situation is schematically shown in Fig. 8.5 for the case of two domains (top) and three domains (bottom).

This motion of the domains can be very nicely traced in the current–voltage curves. Figure 8.6 shows the data for two different superlattices [20]. One trace is for a nominally undoped n^+in^+ sample (a), and the other is for a doped n^+nn^+ structure (b). The undoped sample displays a rather smooth current–voltage curve, whereas the doped sample shows many discontinuities

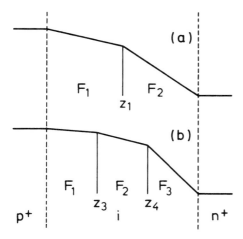

Fig. 8.5. Potential in a sample with two domains (*top*) and three domains (*bottom*) (from [20])

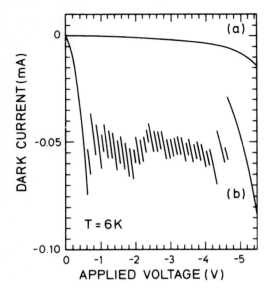

Fig. 8.6. Current–voltage curves at $T = 6$ K for a nominally undoped n^+in^+ sample (**a**) and for a doped n^+nn^+ sample (**b**) (from [20])

in the curve. Each of these discontinuities corresponds to one "hop" of the domain boundary to an adjacent well. Therefore, one can roughly count the number of wells in the sample (in this case, 40).

8.3 High-Field Transport: Observation of Negative Differential Conductivity

From the theoretical considerations outlined above, it might sound comparatively easy to observe high-field phenomena, including negative differential conductivity, in current–voltage curves. However, there are some problems with such an experiment:

- In many cases, the electric field is not homogeneous, and regions of different field contribute in different manners to the current–voltage curve.
- In the experiments, the superlattice region must be embedded between (bulk) contacts. The transition between these bulk regions and the superlattice creates additional obstacles to the transport which can mask the superlattice effects that one is looking for. This problem can be eased by including suitably designed graded structures between the contacts and the superlattice structure.
- In general, conductors with negative differential conductivity tend to be unstable: a local statistical increase in the electron current will lead to a field reduction for the electrons concerned. When they are in the negative-differential-velocity regime, their speed will thus increase and the perturbation will build up.

8.3 High-Field Transport: Negative Differential Conductivity

For these reasons, it has taken surprisingly long after the invention of the superlattice for its unusual high-field transport properties to be convincingly established. In the following, we discuss experiments on undoped superlattice structures first, and then experiments on doped structures.

8.3.1 Experiments on Undoped Superlattices

We describe here experiments by Sibille et al. [21, 22], where the current–voltage curve of an undoped 35.5/20 Å GaAs/AlAs superlattice sandwiched between GaAs n^+ contacting layers was measured. Because of the difficulties of interpretation discussed above, a full self-consistent solution of the coupled transport and Poisson equations was performed and compared with experiment. As a further component, the velocity–field relation $v(F)$ had to be parametrized to distinguish between the different possible mechanisms. In the analysis of Sibille et al. [21, 22], the following relation was used:

$$v = \frac{\mu F}{1 + (F/F_\mathrm{C})^\eta} \,, \tag{8.11}$$

which corresponds to the Esaki–Tsu equation (8.5) for $\eta = 2$, a saturation law for $\eta = 1$, and ohmic conduction for $\eta = 0$. Negative differential velocity is present for $\eta > 1$.

The dots in Fig. 8.7 are the experimental data. For modeling, a number of parameters have to be known or obtained from a fit: the low-

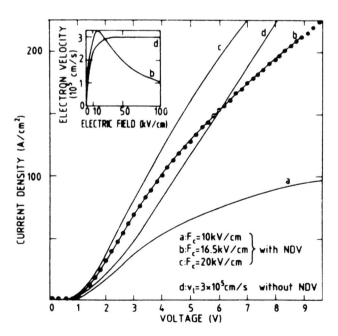

Fig. 8.7. Current–voltage curve for a superlattice structure. *Dots*: Experimental data. *Solid lines*: theory, as described in the text. Line (b) is based on the Esaki–Tsu relation (from [22])

field mobility μ, the critical field F_C, the current–voltage curve parameter η, and the residual acceptor density N_A in the superlattice.[1] Despite the fact that four parameters were unknown, it turned out [1, 22] that the fits are significant, owing to the very different action of each parameter on the current–voltage curve. Accordingly, the solid lines in Fig. 8.7 show that an excellent fit to the experimental current–voltage relation is obtained if the Esaki–Tsu equation $\eta = 2$ is assumed and a critical field of $F_0 = 16.5$ kV/cm is chosen (line (b)). Lines (c) and (d) show Esaki–Tsu relations with different critical fields, and line (a) models a saturation law with $\eta = 1$. Further parameters obtained from the best fit (b) were a low-field mobility $\mu = 40$ cm^2/(V s) and a residual acceptor density $N_A = 1.4 \times 10^{15}$ cm^{-3}, which is typical for such structures. The fit method was also validated by modeling bulk GaAs structures, where the parameters are well known.

8.3.2 Experiments on Doped Superlattices

The larger amount of mobile charges in doped superlattices requires a careful modeling of the coupled transport and Poisson equations. A detailed analysis [24] shows that instabilities can indeed occur and that domains form in the sample. Accordingly, Fig. 8.8 can be interpreted that in the lower left panel a high-field domain forms at the anode side of the superlattice sample. Nevertheless, the model for $v(F)$ based on the Esaki–Tsu theory describes the current–voltage curves for different samples quite well (Fig. 8.8, right), thus confirming the result from the undoped samples that the current–voltage curves can be described using the Esaki–Tsu theory (8.5).

However, the data shown in Fig. 8.8 also demonstrate the subtleties of the electrical transport experiment: owing to the unstable nature of the transport in the superlattice beyond the critical field, even small variations in the way the sample is contacted change the electrical behavior qualitatively.

8.3.3 Magnetotransport Experiments on Superlattices

In semiconductor superlattices, with their pronounced dimensionality effects, the influence of an applied magnetic field is particularly interesting.

If the field is applied parallel to the superlattice layers (i.e. perpendicular to the electric field), the electrons which are transported in the z direction additionally perform a kind of a cyclotron motion in the xz plane. Depending on the relation between the electric and the magnetic field, both closed and open orbits can be obtained. A number of studies have addressed this case [25–29].

---

[1] Nominally undoped AlGaAs/GaAs heterostructures are usually weakly p-doped owing to residual acceptors.

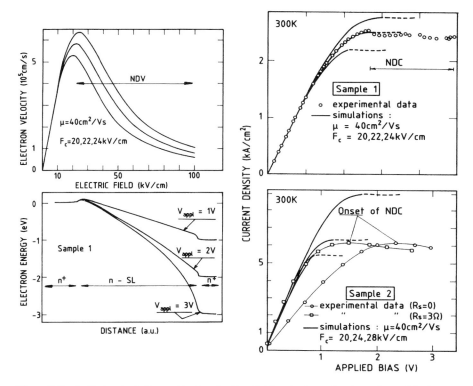

Fig. 8.8. *Left*: typical theoretical velocity–field relations (*top*) and modeled conduction band diagrams under bias (*bottom*) in weakly doped n^+nn^+ superlattice structures. *Right*: experimental current–voltage curves of two n^+nn^+ GaAs/AlAs superlattices *symbols*), together with the modeled (*thick lines*) characteristics (from [23])

If the magnetic field is applied parallel to the electric field, the carriers perform a cyclotron motion in the xy plane. For a sufficiently high magnetic field, Landau quantization then leads to a true one-dimensional system. The dispersion of the system in the quantum limit is given by

$$E(\mathbf{k}) = E_z(k_z) + (n + 1/2)\hbar\omega_C , \quad (8.12)$$

where $\hbar\omega_C$ is the cyclotron energy. The basic principle is that if the cyclotron energy is larger than the width of the first miniband, then the electron will only be able to gain energy along the superlattice axis.

Noguchi and Sakaki [30] have investigated transport for parallel electric and magnetic fields. In particular, they considered the case where the cyclotron energy and the longitudinal optical-phonon energy are in resonance, i.e. $n\hbar\omega_C = \hbar\omega_{LO}$.

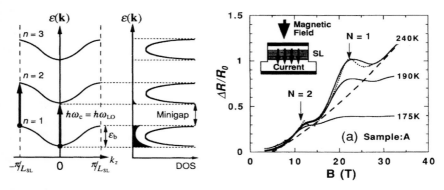

Fig. 8.9. *Left*: dispersion and density of states of a superlattice in a high magnetic field. *Right*: longitudinal magnetoresistance of a superlattice at different temperatures showing pronounced magnetophonon resonances (from [30])

The left panel of Fig. 8.9 displays the dispersion. The resonances correspond to a scattering of electrons from the lowest Landau level to a higher level, for example. The right panel of Fig. 8.9 shows the magnetotransport data. When the resistance is measured as a function of magnetic field, the resonances are clearly observed as maxima in the resistance. The effect is quite pronounced, owing to the one-dimensional nature of the system.

8.3.4 AC Conductivity Measurements of Superlattice Transport

A device which is dominated by linear transport is not expected to show a pronounced response to an AC electric field. However, for nonlinear transport, a pronounced interaction between the DC background transport and the AC conductivity effects is expected. In simple words, the external AC field modulates the local static DC field, which raises or lowers the mobility of the carriers owing to the essence of nonlinear transport, the dependence of mobility on field.

An effect rather similar to effects in superlattices is the Gunn effect in bulk GaAs, where the field is high enough that the carriers are transferred from the light-mass central valley of the conduction band to a higher, heavy-mass valley. It has been shown that the small-signal AC transport can be analytically calculated for the Gunn effect [31]. These results can be applied qualitatively to our case. The calculations predict resonances in the conductance when the AC frequency ν is an integer multiple of the reciprocal ensemble average transit time through the sample, of thickness L:

$$\langle v \rangle / L = n \nu \,. \tag{8.13}$$

These predictions were investigated in detail by Sibille et al. [32, 33] in a series of experiments. The results (see Figs. 8.10, 8.11, and 8.12) agree

8.3 High-Field Transport: Negative Differential Conductivity   165

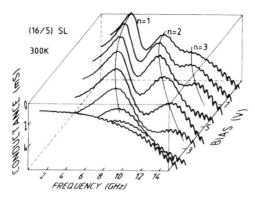

Fig. 8.10. Bias voltage dependence of the AC conductance spectrum in an undoped 4.5 nm/1.4 nm GaAs/AlAs superlattice (from [22])

very well with the qualitative arguments given above. The main points visible are:

- With increasing bias, the AC resonance peak positions shift to lower frequency, confirming the existence of negative differential velocity (Fig. 8.10).

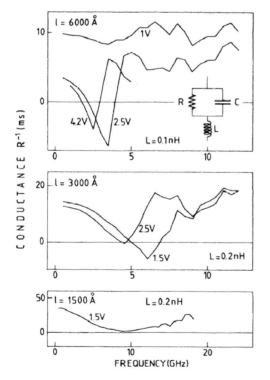

Fig. 8.11. AC conductance of three samples with identical superlattice structures, but different thicknesses L, displaying the scaling of the resonance (minima) with inverse thickness (from [32]

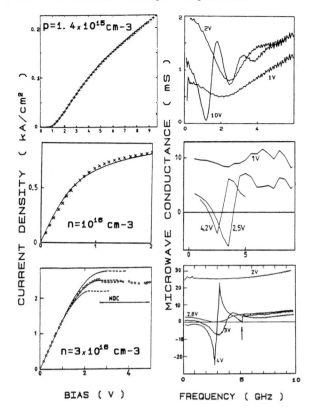

Fig. 8.12. Current–voltage characteristics and AC conductance spectra for a series of GaAs/AlAs superlattices with increasing doping (from [33])

- The inverse scaling of the resonances with superlattice thickness (8.13) is nicely reproduced (Fig. 8.11).
- With increasing doping density, the transport becomes more and more sublinear, and finally shows negative differential conductance (Fig. 8.12, left panel).
- Similarly, with increasing doping density, the AC field resonances sharpen and their amplitude increases (Fig. 8.12, right panel).

8.3.5 Superlattice Transport Measurements by Time-of-Flight Techniques

By detection of the currents of optically generated carriers, it is possible to directly observe the transit behavior of carriers through a sample in the time domain. This "time-of-flight" technique, which consists of the generation of an electron–hole sheet in the sample by a pulsed optical source and the electrical detection of the subsequent transport, has several advantages over simple transport experiments:

8.3 High-Field Transport: Negative Differential Conductivity

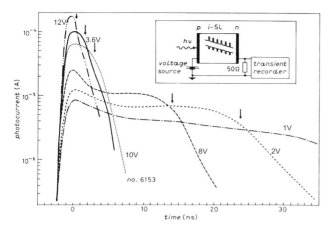

Fig. 8.13. Transient photocurrent at 100 K for different applied reverse voltages. The arrows mark the values which were evaluated as the transit times (from [34])

- By suitable design of the sample and choice of the excitation wavelength, it is possible to generate the carriers at a well-defined location in the sample.
- Electrons and holes can be separated because generation at one side of the sample allows only one type of carrier to contribute significantly to the current.
- The pulsed nature of the experiments prevents field instabilities in the regime of negative differential velocity.

In an initial study by Schneider et al. [34], the photocurrent in a 50 period 128/20 Å GaAs/AlAs superlattice was investigated as a function of bias voltage. Those authors investigated the photocurrent under pulsed excitation with a nitrogen-pumped dye laser with a 500 ps pulse duration, both with time-resolved detection and with time-integrated detection.

Figure 8.13 shows the transient photocurrent traces, which nicely show the nonmonotonic behavior of the transit time through the superlattice. The transit time constants derived from these data are displayed in Fig. 8.14. The transit time initially quickly decreases to a minimum at about 3.6 V, increases again, until it decreases again and reaches a second minimum at 11 V. The authors of [34] showed that the minima are caused by resonant Zener tunneling from the first miniband to the second (3.6 V) and the third (11 V) miniband state in the adjacent well. However, a precise determination of the tunneling times is not possible, owing to inhomogeneous broadening of the tunneling resonances.

Figure 8.15 shows the peak photocurrent vs. the applied voltage. The photocurrent nicely shows the same resonances as those observed in the transit time.

The behavior in the low-field regime, which is shown in the inset of Fig. 8.15 on a linear scale, is quite interesting. Here, the sample reaches flat band around 1.5 V forward bias, as indicated by the zero of the photocurrent. With decreasing forward bias, the current first increases linearly, but then

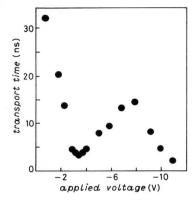

Fig. 8.14. Transit time vs. applied voltage (from [34])

saturates and decreases. The data thus clearly show the negative differential conductivity due to field-induced localization. Since the optically generated carrier concentration is constant, one can, at least indirectly, also conclude negative differential velocity.

We now discuss another study, by Le Person et al. [35], of negative differential velocity in superlattices. Figure 8.16 shows the results of a time-of-flight experiment on a GaAs/AlAs superlattice sample with a 22 meV width of the first miniband. The peak observed at a fraction of a nanosecond in the experimental data (left panel) is ascribed to the photogenerated carriers. This peak is followed by a long-lasting second contribution, which is due to photoinduced electron injection from the cathode.

It is obvious that the arrival time of the photogenerated carriers increases with field, showing the negative differential velocity unambiguously. The theoretically modeled data (right panel) confirm this result very well.

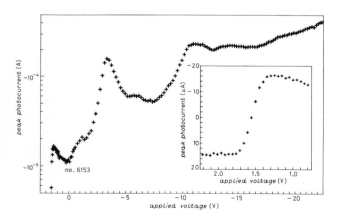

Fig. 8.15. Peak photocurrent as a function of applied voltage at 80 K. The *inset* shows the photocurrent around zero bias on a linear scale (from [34])

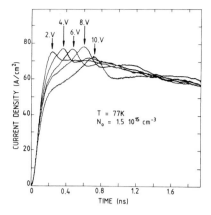 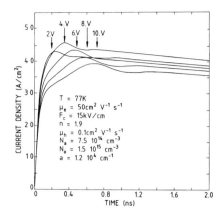

Fig. 8.16. Experimental (*left panel*) and modeled (*right panel*) time-resolved photocurrents in a superlattice. The shift of the main peak to longer delay times with increasing bias voltage is clearly observable (from [35])

8.4 Details of Miniband Transport

8.4.1 Introduction

In the previous sections, we have shown that various experiments have confirmed the basic Esaki–Tsu theory quite nicely. In the following, we shall address some more detailed phenomena where one has to go beyond the simple model to describe all the effects involved. First, we shall address the influence of the coupling between the miniband transport in the z direction and the energy dispersion in the xy directions. Then, we shall address the transport in narrow minibands, in particular under the influence of disorder.

8.4.2 Coupling of Miniband Transport to the xy Dispersion

The Esaki–Tsu model and the solution of the Boltzmann equation in the relaxation time approximation are strictly one-dimensional models. Real superlattice structures are three-dimensional, with dispersion both in the miniband z direction and in the xy plane. Scattering processes will clearly couple the two dispersions, i.e. there will be carrier heating in the xy directions when the carriers are accelerated in the z direction. One might expect that the heating in the xy plane would have a significant impact on the velocity–field relation, since the carrier heating, which strongly influences the $v(F)$ relation (8.10), will be changed when an additional dispersion in the two other directions is present.

Figure 8.17 shows a comparison between experimental data [37] and a theoretical calculation for a three-dimensional system [36]. The most significant

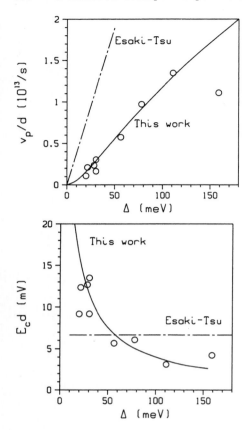

Fig. 8.17. Dependence of peak velocity $v_{\max}$ divided by the superlattice period d (here denoted as v_p/d) and critical field times superlattice period $F_C d$ (here denoted as $E_C d$) on the miniband width Δ. *Solid lines*, theory; *open circles*, experimental data from [22]. The *dashed lines* were obtained from the Esaki–Tsu theory using $\tau = 0.1$ ps (from [36])

result is the decrease of the critical field with increasing miniband width Δ, which cannot be understood in the Esaki–Tsu theory. The marked differences between the Esaki–Tsu theory and the three-dimensional model are obvious from the data in Fig. 8.17.[2] The basic effects are also clearly visible in the data of Fig. 8.17.

This surprising decrease of the critical field F_C has been explained by Gerhardts [38], who has shown that it is a direct consequence of the interplay of elastic and inelastic scattering: a growing importance of the former process relative to the latter increases lateral electron heating. Since this is increased for higher electric fields, there is a stronger reduction of the velocity owing to lateral heating (compared with the Esaki–Tsu model) for large fields than for small fields. Thus, F_C is shifted to lower fields for wide minibands.

---

[2]Despite the excellent modeling of the experimental data, note the reservations raised against the theoretical approach chosen here [1].

8.4.3 Transport in Narrow Minibands: Suppression of Phonon Scattering and Influence of Disorder

In the following, we study the transport in narrow minibands, where the influence of phonon scattering and disorder is particularly important.

It is quite interesting to study the transport in narrow minibands for the following reasons:

- If the miniband width becomes smaller than the optical-phonon energy, the phonon scattering rate will be greatly reduced, since only acoustic-phonon scattering is possible. This process is slow compared with the impurity and interface roughness scattering.
- For narrow minibands, the influence of fluctuations in the well and barrier widths should become stronger. This will lead to an increasing tendency to localization of the Bloch waves in the minibands.

Suppression of Phonon Scattering The first detailed study of miniband transport in weakly coupled superlattices with narrow minibands was performed by Grahn et al. on a series of superlattices with miniband widths of 0.5 to 1.5 meV, using time-of-flight experiments [39].

Figure 8.18 shows the velocities obtained from the time-of-flight experiment. The negative differential velocity at high fields is clearly discernible. It was observed up to sample temperatures of 130 K, well above the temperature where $k_B T$ becomes larger than the miniband width.

Further studies on narrow minibands were performed by Sibille et al. [40] on a series of weakly n-doped GaAs/AlAs superlattices with miniband widths of 4 and 16 meV. The current–voltage data and the velocity–field relations obtained by modeling based on the Boltzmann equation are shown in Fig. 8.19. The data reveal both negative differential conductivity and negative differential velocity up to room temperature.

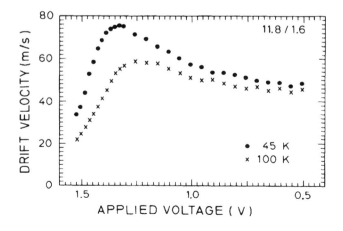

Fig. 8.18. Drift velocity as a function of bias voltage for a 11.8/1.6 nm GaAs/AlGaAs superlattice (miniband width ≈ 1.5 meV) at lattice temperature of 45 and 100 K (from [39])

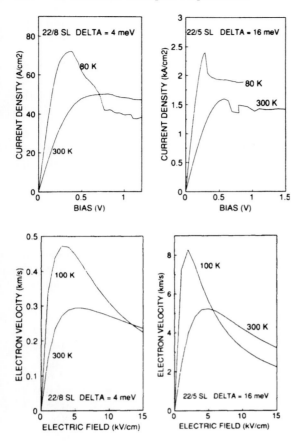

Fig. 8.19. *Top*: current–voltage characteristics of narrow-miniband superlattices with first-miniband band widths of 4 and 16 meV. *Bottom*: velocity-field relations obtained by a Boltzmann model for the same samples (from [40])

Sibille et al. [40] have also studied the temperature dependence of the current–voltage curves. With increasing temperature, the peak current densities decrease. The same holds for the peak velocity obtained from the model. At the same time, the peak conductivity and the peak velocity shifting towards higher fields when the temperature is increased. The experimental observations are summarized in Fig. 8.20.

The low-field conductance and thus the low-field mobility follows the relaxation time approximation rather well. This could indicate that the scattering processes in a narrow miniband are mostly elastic, which is a key factor for the applicability of the relaxation time approximation. The most likely candidate for the elastic, temperature-independent scattering term is the interface roughness scattering.

In Fig. 8.21, the parameters μ/d^2, V_p/d, and edF_C are summarized as a function of miniband width for a large number of GaAs/AlAs superlattices, with miniband widths Δ ranging from 4 to 130 meV. Both the low-field

8.4 Details of Miniband Transport   173

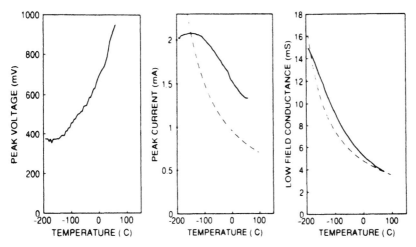

Fig. 8.20. Temperature dependence of the current–voltage characteristics (*solid lines*) of a superlattice with 4 meV miniband width. The data reasonably agree with the Esaki–Tsu theory (*dashed lines*) for the low field mobility (from [40])

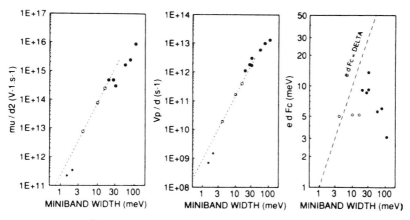

Fig. 8.21. μ/d^2, V_p/d, and edF_C as a function of miniband width (V_p is the peak velocity which is denoted as v_max in this book). The data are from [40] (*open circles* and *full circles*) and [39] (*stars*, lowest two data points in left and center panel)

mobility and the peak velocity follow a quadratic dependence on Δ over several orders of magnitude.

Influence of Disorder All the theoretical models which we have discussed so far are based on delocalized electron states, i.e. Bloch waves. For weak perturbations, one can treat their effects in perturbation theory and assume that the Bloch waves are preserved. However, for strong perturbations, for example by disorder, this fails and the Bloch waves are no longer suitable eigenstates for describing the problem. Instead, the band structure separates into localized states.

A qualitative argument to describe this transition is to compare the width of the miniband, as a measure of the coupling strength of the superlattice, with the inverse of the scattering time. From the experiments discussed above, typical scattering times are on the order of 100 fs, which corresponds to a lifetime broadening of about 6.5 meV. It is thus obvious that in this range, the very concept of a miniband becomes questionable.

The same argument holds for the broadening induced by static disorder in the superlattice, for example that caused by fluctuations of the well and barrier widths. If the energetic broadening due to this effect exceeds the miniband width, the concept of delocalized Bloch waves is also no longer applicable.

The effect of disorder has been studied in detail by Chomette et al. [41] who studied intentionally disordered GaAs/AlGaAs superlattices by randomly changing the well width of the samples. As a parameter for the disorder, the Gaussian standard deviation σ (denoted in that work by S) around an average well width of 3 nm was chosen.

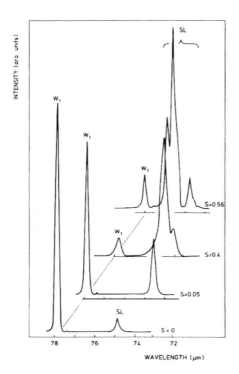

Fig. 8.22. Photoluminescence spectra of four superlattice samples with increasing degree of disorder σ (denoted here by S) (from [41])

8.4 Details of Miniband Transport   175

The vertical transport was studied using the enlarged-well optical technique, as described in Sect. 8.2.1. The vertical transport starts to be strongly suppressed if the standard deviation σ reaches about 0.5 nm; this effect is visible in the suppression of the enlarged-well photoluminescence, compared with the superlattice signal (Fig. 8.22).

Figure 8.23 shows the ratio of the superlattice photoluminescence to the enlarged-well photoluminescence (which is a measure of slow vertical transport), as a function of the disorder parameter σ for various lattice temperatures. With increasing disorder, the transport is suppressed. This phenomenon, however, is obviously dependent on temperature. At low temperature, the suppression of transport starts at a very low disorder; at higher temperatures, the transport is still efficient for higher values of the disorder parameter.

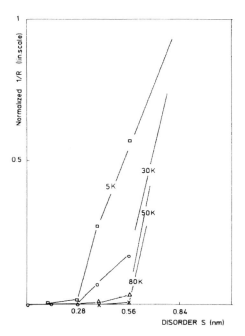

Fig. 8.23. Temperature dependence of the ratio of the superlattice photoluminescence to the enlarged-well photoluminescence as a function of the disorder parameter σ for various lattice temperatures (from [41])

This effect can be well explained by the fact that the transport is dominated by localized states at the lower edge of the band. States further up in energy tend to be more delocalized. Thus, thermal excitation improves transport despite the disorder.

When considering the results of Chomette et al., one has to keep in mind that in superlattices without intentional fluctuations of the well width, the disorder varies laterally. Thus, there is an averaging effect which reduces the importance of disorder. Also, the electrical transport will then be of

8.4.4 Electrical Spectroscopy Using Transistor Structures

Here, we discuss a number of experiments which have used suitably designed structures to perform electrical spectroscopy on the conduction band only. Such experiments are possible in transistor structures where the carriers travel at least partly ballistically.

England et al. [37] where the first to use such an approach. They investigated a unipolar hot-electron transistor. Here, electrons are injected through a barrier with a larger gap into the base region. After traveling through the base, the carriers then reach the collector, which is a second barrier. Gain is observed if the carriers travel ballistically. By raising and lowering the injection energy of the carriers by varying the emitter–base voltage, one can determine the energy spectrum of the electrons after they have traveled through the base.

England et al. [37] used this principle by inserting a superlattice structure into the base. They obtained a series of resonances in the current, which was caused by the different minibands of the superlattice (see Fig. 8.24). The current shows maxima when the injection energy is aligned with a miniband; in between, the carriers experience a quantum reflection, which reduces the current. The low gain values measured correspond to about three periods of the superlattice, which leads to scattering times of 20–50 fs, somewhat below what later measurements have yielded.

Recently, the group of Gornik et al. has performed a number of elegant experiments on the transport in superlattices. In these experiments, a modified resonant-tunneling transistor was used (see Fig. 8.25). The sample structure consisted of a resonant-tunneling injector (producing a rather narrow electron distribution in energy space), followed by a base structure and a collector.

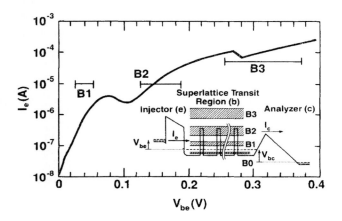

Fig. 8.24. Injection current in a hot-electron transistor, with a 120/25 Å $GaAs/Al_{0.3}Ga_{0.7}As$ superlattice in the base (from [37])

The key point is that the collector incorporated a superlattice structure, similarly to the approach of Beltram et al. [42]. By measuring the collector current as a function of the base-collector voltage, it was possible to follow the transport of the ballistically injected electrons through the superlattice.

The experiments clearly showed the miniband resonances and the minigaps in the superlattice structure for low fields. For higher fields, the Wannier–Stark ladder became obvious as resonances in the current [43]. Figure 8.26 shows the miniband resonances in the transfer ratio of the collector. The data clearly show the miniband resonances, albeit with some broadening due to the energy distribution of the incoming beam. Also, phonon replicas are visible, since some incoming electrons have been scattered by LO phonons.

In subsequent papers, coherent and incoherent transport was investigated [44,45]. The approach was to vary the number of periods in the superlattice from 5 to 30 periods. For the five-period structure, the transmission was symmetric when positive and negative biases were applied to the superlattice structure. For the structures with a larger number of periods, the transmission became asymmetric because incoherent, dissipative transport was also taking place. From comparison with theory, the authors of [44, 45] concluded that interface scattering was responsible for the loss of coherence, rather than phonon scattering.

Figure 8.27 shows the results for the 20-period sample, taken at 4.2 K. The total transmission (dots) is asymmetric owing to an incoherent contribution. The incoherent current (open squares) was obtained by comparison with theory. This current clearly shows negative differential conductivity and corresponds very nicely to the Esaki–Tsu theory, assuming a scattering time τ of 1 ps.

Note that in the experiments just discussed, the electrons entering the superlattice do not form a wave packet, since a continuous beam of electrons is injected and their homogeneous width is quite small. The total spectral

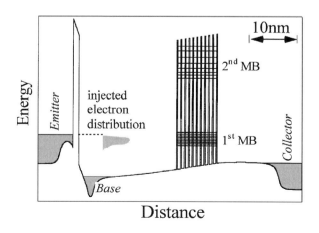

Fig. 8.25. Schematic band diagram of a resonant-tunneling transistor device, with a negative bias applied to the superlattice (from [44])

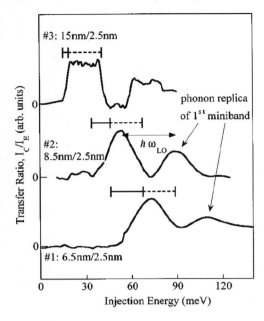

Fig. 8.26. Transfer ratio through the superlattice vs. injection energy at lower injection energies for all three samples used. The *solid bars* indicate the calculated miniband position; the *dashed bars* indicate the broadening due to the energy distribution of the injected electron beam. The *double arrow* represents the energy of a longitudinal optical phonon (from [43])

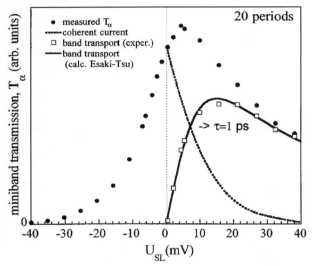

Fig. 8.27. Miniband transmission vs. bias for the 20-period superlattice (*dots*). For positive applied bias, the coherent part of the transmission (*dashed line*) was subtracted to yield the incoherent part (*squares*). The incoherent current is well described by the Esaki–Tsu theory (*solid line*) assuming a scattering time of $\tau = 1$ ps (from [44])

width of the electron beam which was injected into the base was actually quite large and on the order of 20 meV. However, this width was caused by inhomogeneous broadening of the injection, not by the intrinsic homogeneous width of the wave packets injected into the superlattice. Therefore, these experiments do not address coherent wave packets, unlike the optical studies. It would be quite interesting if the experimental approach used here could be extended to the generation of very short electron pulses which were broad enough to study electron wave packet propagation.

Relation of Electrical Transport to Bloch Oscillations The very existence of the Wannier–Stark ladder in the optical spectra of superlattices shows also that some of the carriers of the ensemble must perform Bloch oscillations. This follows from the scattering-time argument in Sect. 1.3, but also intuitively: Bloch oscillations occur when the carriers are accelerated in the electric field till they reach the upper edge of the band, "tilted" by the electric field. This, however, is obviously equivalent to the condition that the coherence length of the electron is identical to the field localization length Δ/eF. The electron wave functions thus "feel the edges" of the band and therefore display a discrete spectrum. Owing to the ensemble nature of the electrical transport, only some of the carriers will perform full Bloch cycles. In fact, as argued previously, if all carriers were to Bloch oscillate (i.e. $\tau \to \infty$), the ensemble average velocity would be zero. Transport is thus only possible owing to the very existence of scattering. Note, however, that these arguments are only fully applicable if only one band is present. The coupling to higher bands changes this entirely, as has been argued in Chap. 7.

8.5 Negative Differential Conductivity in Bulk Silicon Carbide

As discussed previously, the observation of Bloch oscillations in bulk materials is difficult, owing to the wide band widths and the very high fields needed. An intermediate scenario between bulk materials and artificial superlattice structures is provided by bulk materials which have a very large unit cell, at least in one crystallographic direction. One of those materials is the semiconductor silicon carbide, SiC, which has a variety of different polytypes. Figure 8.28 shows a section of the SiC crystal in the [11$\bar{2}$0] plane. There are basically two packing schemes in the direction of the c axis of the crystal: the 3C packing scheme uses the three different places A, B, and C in the hexagonal closest-packing scheme in sequence; the 2H packing scheme uses only the A and B places. By combining the two packing schemes, it is possible to generate a large number of polytypes with different repeat lengths. The polytypes 4H, 6H, and 8H, which have unit cell lengths in the stacking direction of 5 Å, 7.5 Å, and 10 Å, respectively, are shown schematically in Fig. 8.28.

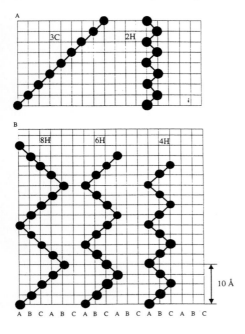

Fig. 8.28. *Upper part* (A): arrangement of the stacking in SiC for the types 3C and H. *Lower part* (B): stacking arrangements for the polytypes 4H, 6H, and 8H (from [49])

The band gap of the 2H variant is about 1 eV larger than that of the 3C variant. The 4H, 6H, and 8H polytypes can therefore be seen as natural superlattices in the direction of the c axis, which consist of alternating wells (the regions of 3C type) and barriers (the 2H "kinks").

The SiC system was investigated in detail by Sankin and coworkers in a sequence of studies using high-field transport experiments [46–49]. These authors observed current–voltage curves with negative differential conductivity if the electric field was parallel to the c axis. In the most recent paper [49], it was conclusively shown that this effect is indeed caused by Bloch oscillations. Here, the authors compared the 4H, 6H, and 8H polytypes of SiC and showed that the critical field scales with the period d as expected.

The current–field relation for the 8H structure at room temperature is shown in Fig. 8.29. The electrical measurements were performed with an elaborate technique, needed to achieve homogeneous fields and to avoid disturbing effects such as the influence of hole transport. The data clearly show a pronounced decrease of the current and thus negative differential conductivity once a certain threshold field is reached. It is observed experimentally that the threshold field scales with $1/d$, as theoretically expected. For instance, the width of the first miniband in the 6H structure is estimated to be 100–400 meV; the width of the gap between the first and second miniband is estimated to be 300–500 meV [50].

The scattering time τ was evaluated from the data using a theoretical model [49]. Scattering times of 175 fs, 205 fs, and 110 fs were obtained for

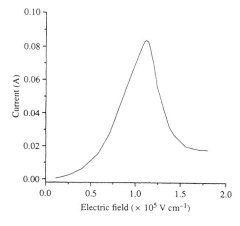

Fig. 8.29. Current–field relation for the 8H polytype (from [49])

the 4, 6, and 8H modifications, respectively. These numbers are actually quite close to the values obtained for GaAs/AlGaAs superlattices.

8.6 Combined Transport and Optical Experiments

8.6.1 Comparison of Transport and Optical Experiments

In Chaps. 3 to 5, we have used only the Wannier–Stark ladder picture to describe the system. For the discussion of the transport experiments in this chapter, however, we have relied on a description on the basis of the $F = 0$ Bloch wave states. It is interesting to discuss how these two descriptions correspond.

First, we shall discuss a comparison of theoretical models based on localized Wannier–Stark levels with the theoretical model presented above, based on delocalized states. Then, we shall discuss key experiments by Sibille et al. which directly prove that the occurrence of negative differential velocity is linked to the existence of the Wannier–Stark ladder. Finally, we discuss the possibility that a similar relation also holds for negative differential velocity and Bloch oscillations.

Hopping Transport vs. Band Transport An alternative way to regard band transport in a miniband at high fields is to consider it as hopping transport between localized states. Such a picture was first developed by Tsu and Döhler [51]. The theory used acoustic-phonon scattering as the only scattering mechanism. A further simplification was that the model assumed Fermi–Dirac distributions in the states of the Wannier–Stark ladder, i.e. the generation of nonequilibrium carriers by the hopping process was assumed to be a perturbation.

Figure 8.30 shows the current in the Wannier–Stark ladder as a function of the electric field. For low fields, the calculations predict a conductance linear with the field; for high fields, a maximum followed by negative differential conductivity is predicted. It is interesting to note that the theory predicts an increase of the current with temperature due to the increased hopping rate, a result which is opposite to that of miniband transport models discussed above.

It is quite obvious that the hopping model will be inaccurate at very low fields when the picture of band transport is fully applicable. However, it is interesting to discuss at what field the hopping model becomes realistic. The key criterion is that scattering increases the transport instead of decreasing it. In the Esaki–Tsu model, this field can be easily determined by calculating the derivative $dv/d\tau$ of the ensemble velocity with respect to the scattering time. It turns out [1] that the derivative is positive if $F < F_C$ and negative if $F > F_C$. This means that scattering inhibits transport if the field is below the critical field (which is the typical behavior for band transport) and scattering increases transport above the critical field (which is the typical behavior for hopping transport). It is thus obvious that the hopping transport model is a meaningful description for transport at fields above the critical field.

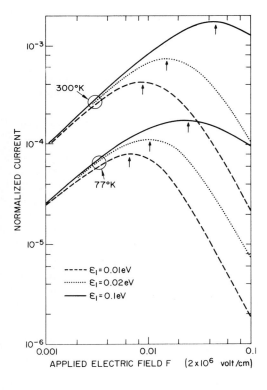

Fig. 8.30. Dependence of current on electric field for the tunneling transport between Wannier–Stark states for minibands with different widths of the first miniband (from [51])

8.6 Combined Transport and Optical Experiments

The availability of a large amount of experimental results starting around 1990 has stimulated more theoretical work to clarify the relation between Esaki–Tsu-like models and other approaches (see, e.g., [52–59]). Wacker and Jauho [56] used a nonequilibrium Green function theory to directly compare the different approaches.

Figure 8.31 shows a sketch of the validity ranges of the theories taken from [56]. The data are given in terms of the electric field, expressed in terms of the field energy eFd, vs. the transfer integral, expressed here as T_1 (the miniband width Δ is four times T_1).[3] Generally, one can conclude that the sequential-tunneling approach is only valid for narrow-band superlattices which are much narrower than the energy width associated with the scattering rate Γ. The miniband conduction theory and the hopping theory hold for larger band widths. The miniband theory is valid for lower fields, while the hopping theory addresses higher fields.

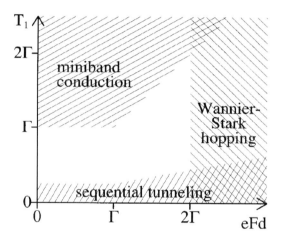

Fig. 8.31. Validity ranges of different theoretical models. Data are given in terms of the electric field, expressed as eFd, vs. the transfer integral, expressed as T_1 (miniband width $\Delta = 4T_1$). The quantities are all given in units of the scattering rate $\Gamma = 1/\tau$ (from [56])

These considerations are shown more quantitatively in Fig. 8.32, which displays calculations of the current–voltage curves for the different theories, compared with the results of the "true" nonequilibrium Green function approach, marked NGF. The calculations show nicely that the miniband conduction theory (Esaki–Tsu and comparable models) works well for lower fields, but fails for high fields when the field energy exceeds the transfer integral. In this regime, the Wannier–Stark hopping theory and the sequential-tunneling theory match better to the NGF approach.

While the calculations by Wacker et al. [56, 57] show nicely the applicability of the different models, they were not intended to include all scattering

---

[3] The notation used here for the transfer integral (T_1) should not be confused with its more frequent use for the population lifetime.

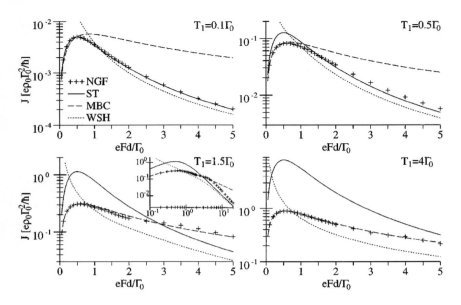

Fig. 8.32. Current–voltage curves calculated using various theoretical approaches: nonequilibrium Green functions (NGF, *symbols*), miniband conduction (MBC, *dashed lines*), hopping in the Wannier–Stark ladder (WSH, *dotted lines*), and sequential tunneling (ST, *solid lines*). The electric field is expressed in units of the field energy eFd. The four panels show four different ratios between the transfer integral (i.e. 1/4 of the miniband width) and the scattering width (from [56])

effects in detail so as to enable a quantitative comparison with experiment. Such calculations have been performed using the hopping model by Rott et al. [54,58,59]. The main intention of that work was to show that the hopping model can be extended if not only the interaction with the neighboring wells but also that with all other wells is taken into account. In this case, it can be shown that the hopping theory is an accurate description as long as the Bloch frequency ω_B is larger than the scattering rate $1/\tau$ (which is the condition for the onset of negative differential velocity). This fact can also be well understood by the arguments we have mentioned several times before in this book: for the description of the physics, it should not matter if one uses the eigenstates with a field (Wannier–Stark states) or the eigenstates without a field (miniband states). As long as a complete basis is used, the same results should be obtained.

In their most recent publication [59], Rott et al. calculated the drift velocities in strongly coupled superlattices, including all bands and including elastic (ionized-impurity) and inelastic (acoustic-phonon scattering and polar-optical scattering of LO phonons) scattering processes. Figure 8.33 shows the contributions of the transitions to the neighboring Wannier–Stark state

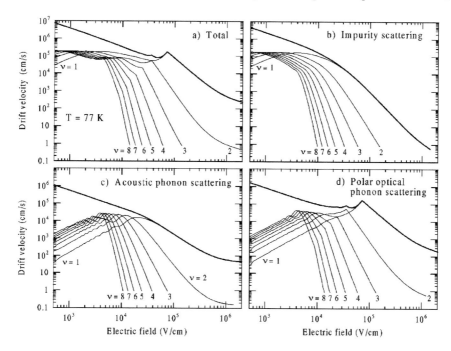

Fig. 8.33. Contributions of the transitions $|0\rangle$ to $|\nu\rangle$ (*thin lines*) to the total drift velocity (*thick line*). $|0\rangle$ is the Wannier–Stark ladder state where the carriers originated, and $|\nu\rangle$ the state where they are scattered to. The four different panels show the results including all scattering processes (**a**) and for individual processes (**b**)–(**d**) (from [59])

$\nu = 1$ and to states further away to the total drift velocity for the various scattering processes and for the total rate. At higher fields, the nearest-neighbor transitions dominate; at lower fields, transitions to states further away are the major contribution.

The calculations also nicely show the influence of phonon resonances and of Zener resonances due to coupling to higher minibands. Figure 8.34 shows the drift velocity as a function of the electric field for various temperatures. The drift velocity shows resonances due to scattering to the nearest neighbor ($eFd = \hbar\omega_{\mathrm{LO}}$) and to the next nearest neighbor well ($2eFd = \hbar\omega_{\mathrm{LO}}$).

Figure 8.35 shows the drift velocity as a function of the electric field, comparing a single-miniband approximation based on Kane functions and the model of Rott et al. [59]. The data of the full model show pronounced resonances when the field energy eFd is equal to the splitting between the first and the second miniband or to half of this splitting.

Experiments on the Correspondence Between the Wannier–Stark Ladder and Negative Differential Velocity It is a corollary of the

186   8 Electrical Transport Experiments

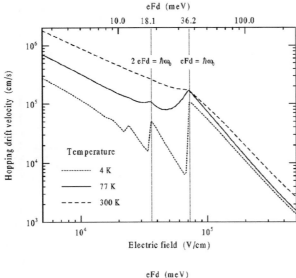

Fig. 8.34. Drift velocity as a function of the electric field for different temperatures. The data show clear resonances caused by optical-phonon scattering to the nearest-neighbor and to the next-nearest-neighbor (from [59])

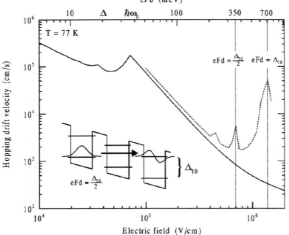

Fig. 8.35. Drift velocity as a function of the electric field, comparing a single-miniband approximation based on Kane functions (*solid line*) and the model of Rott et al. [59] (*dotted line*). Resonances occur when the field energy eFd is equal to the splitting between the first and the second miniband denoted by Δ_{10} here (nearest-neighbor Zener tunneling), or to half of it (next-nearest-neighbor Zener tunneling) (from [59])

discussion above that the region of fields above the critical field should be characterized by the existence of the localized states of the Wannier–Stark ladder. Therefore, it is of great importance to prove experimentally that the

8.6 Combined Transport and Optical Experiments

existence of negative differential velocity is directly linked to the existence of the Wannier–Stark ladder states.

The first such experiment which proved this correspondence was performed by Beltram et al. [42] using bipolar transistor structures with a superlattice placed in the collector. These authors observed negative differential conductivity in the collector (Fig. 8.36), followed by several resonances under high collector bias, which were interpreted as quantum reflections of the injected electrons by the Wannier–Stark ladder.

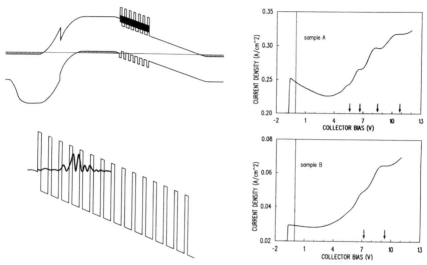

Fig. 8.36. *Left panel*: band diagram of a transistor structure incorporating a superlattice structure in the collector. *Right panel*: collector current density vs. collector bias at fixed emitter current. The superlattice resonances are clearly visible (from [42])

A more direct proof of the correspondence of nonlinear transport and the Wannier–Stark ladder was given by a comparison of transport and optical data. This key experiment was first performed by Sibille et al. [60]. These authors compared conductivity data (both DC and AC) with photocurrent spectra for the same sample (Fig. 8.37). It is obvious from the data that in the field range where negative differential velocity is observed, the transitions of the Wannier–Stark ladder are also present. The experiments thus clearly demonstrate that the results of the high-field transport experiments are consistent with a discretization of the energy spectrum. Therefore, the optical experiments on Bloch oscillations, which can be very well understood as quantum beats of the Wannier–Stark ladder, address the same phenomena as do the high-field electrical transport experiments.

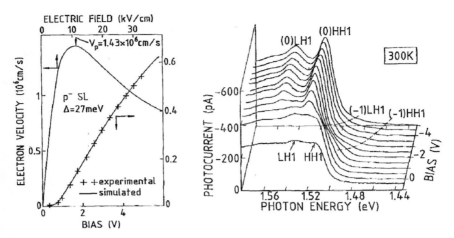

Fig. 8.37. *Left panel*: Modeled electron velocity and current (*solid lines*) and experimentally determined current (*crosses*) of a superlattice, showing negative differential velocity. *Right panel*: photocurrent spectra of the same sample, displaying Wannier–Stark states at the same bias fields (from [60])

8.6.2 Observation of Thermal Saturation in a Superlattice by Infrared Absorption

One important consequence of transport in narrow bands is thermal saturation: if the thermal energy $k_\mathrm{B}T$ starts to approach the band width Δ, the number of carriers in the upper half of the band (with negative mass) becomes comparable to the number of carriers in the lower half of the band (with positive mass). The overall conductivity then becomes suppressed because the number of carriers moving against the field becomes close to the number of carriers moving with the field.

Thermal saturation in the minibands of a semiconductor superlattice was first experimentally demonstrated by Brozak et al. [4, 5]. These authors used far-infrared (FIR) Drude optical-absorption investigations in narrow-miniband GaAs/AlGaAs superlattices ($\Delta = 3$ meV). The experiment required that the electric field of the radiation was oriented along the superlattice growth axis in order to measure the miniband transport instead of the in-plane conductivity. To maximize the coupling between the FIR light and the superlattice, a grating was deposited on the sample surface. A high magnetic field along the z axis removed any parallel transport contribution. Brozak et al. used a 200 Å/20 Å GaAs/Al$_{0.3}$Ga$_{0.7}$As and an 85 Å/10 Å GaAs/AlAs superlattice sample.

Figure 8.38 shows the miniband infrared Drude absorption as a function of temperature. It is obvious that the absorption decreases strongly with temperature, in particular at low frequencies, as expected for thermal saturation.

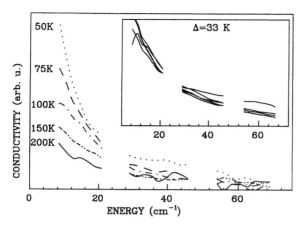

Fig. 8.38. One-dimensional Drude absorption in the miniband of the 200/20 superlattice sample as a function of sample temperature. The *inset* shows the in-plane Drude absorption as a function of temperature. The gaps in the spectra are due to regions of low sensitivity, caused by the beam splitter of the spectrometer (from [5])

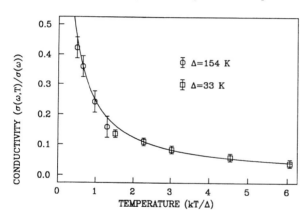

Fig. 8.39. Miniband conductivity at 10 cm^{-1} as a function of temperature for both of the samples (*symbols*) and comparison with theory (*solid line*) (from [5])

The inset of Fig. 8.38 shows the same data for the in-plane xy motion. It is obvious that this absorption is negligibly influenced by the temperature.

Figure 8.39 shows the measured miniband conductivity at 10 cm^{-1} as a function of temperature for both samples. The data are compared with theory (solid line), which predicts

$$\sigma(\omega, T) = \sigma(\omega) \frac{I_1(\Delta/k_\mathrm{B}T)}{I_0(\Delta/k_\mathrm{B}T)} \;, \tag{8.14}$$

where $\sigma(\omega)$ is the usual Drude conductivity and I_1 and I_0 are modified Bessel functions.

The data nicely show that the conductivity approaches zero for $k_\mathrm{B} T \gg \Delta$. The results for both samples are close together and agree with theory very well, confirming that the authors had indeed observed the thermal saturation of the transport.

It should be mentioned here that this study of the intraminiband absorption is only one example of a large number of infrared absorption measurements which have been performed on superlattices to determine the inter- and intraminiband transitions. A detailed overview of this work is given in the review by Helm [61].

8.6.3 Electrical Transport Under Infrared Illumination

It is quite interesting to study the influence of infrared radiation on the electrical transport in superlattices, for the following reason:

- If the photon energy is equal or close to the Wannier–Stark ladder splitting, one can realize an *inverse Bloch oscillator*, i.e. a system where the Bloch oscillating electrons are resonantly driven down the Wannier–Stark ladder.
- If the infrared illumination leads to a change of, for example, the current–voltage curve, one could use a superlattice as a detector.

In the following, we shall discuss some results of such experiments.

In a key experiment, Unterrainer et al. [62] observed the inverse Bloch oscillator predicted by Ignatov et al. [63]. In this study, the current–voltage curves of an 80 Å/20 Å GaAs/Al$_{0.3}$Ga$_{0.7}$As superlattice sample were measured while it was illuminated by intense radiation from a free-electron laser. The current through the superlattice showed resonances when the frequency of the radiation was tuned through multiples of the Bloch oscillation frequency (see Fig. 8.40). These current resonances occur in the voltage range where negative differential velocity is present, i.e. when the carriers perform Bloch oscillations.

Figure 8.41 shows the positions of the steps as a function of the bias voltage across the superlattice. It is obvious that the positions scale linearly with voltage and that there are multiple resonances, as expected from the Shapiro equation

$$eU = nh\nu \ . \tag{8.15}$$

These resonances are an equivalent of the Shapiro steps in superconducting junctions, as discussed in Sect. 4.4.1. They develop because the Wannier–Stark ladder can absorb energy only in multiples of the energy splitting. This infrared experiment thus complements the observation of the Wannier–Stark ladder by interband methods [64, 65].

In another series of experiments, Renk et al. have shown that the electrical behavior of superlattices can be strongly influenced if infrared radiation is focused onto the sample [66–69]. These authors have demonstrated that

8.6 Combined Transport and Optical Experiments   191

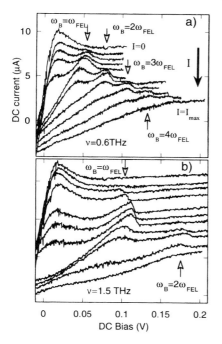

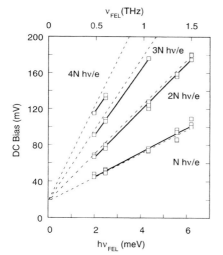

Fig. 8.40. Current–voltage curve as a function of FEL illumination intensity (the curves are shifted downwards for increasing laser intensity). Data for FEL frequencies of 0.6 THz (**a**) and 1.5 THz (**b**) are shown. In the negative-differential-conductivity region, resonances at the Bloch frequency and its subharmonics are visible. The data are taken at 10 K (from [62])

Fig. 8.41. Voltage positions of the Shapiro current peaks versus FEL frequency (from [62])

biased superlattices can thus be used as detectors for FIR radiation [69]. For a discussion of this detector application, see Sect. 10.2.1.

We discuss here an experiment by Winnerl et al. [68] where the change of the current–voltage curve due to irradiation with infrared radiation was investigated. The sample was a 36.3 Å/11.7 Å GaAs/AlAs superlattice which was homogeneously n-doped to a density of 8×10^{16} cm^{-3}. The sample was irradiated with radiation of frequencies 90 GHz, 450 GHz, and 3.6 THz. The irradiation intensity was expressed in the form of the dimensionless parameter $\mu = eE_\omega d/\hbar\omega$, where d is the superlattice period and E_ω is the AC electric-field amplitude.

Figure 8.42 shows the current–voltage curve of the superlattice and the change of the current–voltage curve due to irradiation with 90 GHz and 3.6 THz radiation for various illumination intensities. In all cases, a reduction of the current is observed. For irradiation with 90 GHz radiation, the maximum current reduction is below the critical voltage and shifts with intensity; for irradiation with 3.9 THz, the maximum response is at the critical field and independent of the irradiation intensity.

The influence of infrared radiation on the current–voltage curves was compared with a theoretical model of nonlinear transport in superlattices under illumination. The model can explain well the two regimes. For the 90 and 450 GHz illumination, the response is quasi-static since the relaxation time is much shorter than the AC field cycle ($\omega\tau < 1$). A modulation of the effective field by the illumination will, in simple terms, lead to a negative signal in the regions of the current–voltage curve where the second derivative is negative.

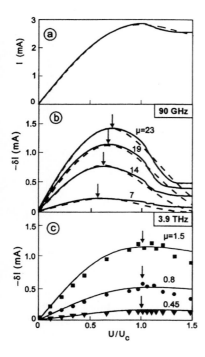

Fig. 8.42. (a) Current–voltage characteristics of a superlattice mesa (*solid line*) and fit with the Esaki–Tsu theory (*dashed line*). (b) Current change for irradiation with 90 GHz radiation: measured (*solid lines*) and calculated for $\omega\tau = 0.4$ (*dashed lines*). (c) Current change for irradiation with 3.9 THz measured (*symbols*) and calculated for $\omega\tau = 3$ (*solid lines*). The parameter μ denotes the infrared field intensity as described in the text (from [67])

For $\omega\tau > 1$, which is the case for the 3.9 THz radiation, the response is dynamic and can be understood in terms of transitions on the Wannier–Stark ladder. When the AC frequency is higher than the Bloch frequency ($\omega > \omega_B$), which is the case for the 3.9 THz illumination investigated here, transitions upwards in the Wannier–Stark ladder (see Fig. 8.43) dominate and the current response is negative. The measurements under infrared illumination can also be used to obtain the momentum relaxation time τ of the carriers. The values at room temperature for the superlattices investigated were approximately 100 fs.

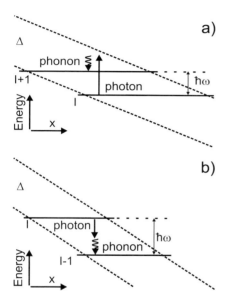

Fig. 8.43. Energy–space scheme for photon absorption (**a**) and photon emission (**b**) transitions on the Wannier–Stark ladder (from [67])

8.6.4 Investigation of Zener Tunneling: Transport and Infrared Experiments

As discussed in Chap. 7, the coupling to higher bands can play a significant role in the coherent motion of the carriers. Here, we discuss recent experiments [70, 71] where the coupling to higher bands was investigated by transport experiments. Such experiments are particularly interesting, since they relate closely to the original work of Zener [72] which triggered the whole research on Bloch oscillations and Zener tunneling.

The first experiment on transport investigations of Zener tunneling was reported by Sibille et al. [70]. These authors investigated the conductivity in intrinsic superlattices embedded in n-doped barriers using small-signal microwave measurements. The intrinsic superlattice structures were chosen to obtain a homogeneous electric field. Figure 8.44 shows the current-voltage curve and the derivative of the conductance of a 99 Å/31 Å GaAs/$Al_{0.25}Ga_{0.75}As$ sample. At higher voltages, the derivative of the conductance shows pronounced resonances. To obtain the velocity–field relations, the experimental data were compared with a theoretical model which used a numerical solution of the Poisson and transport equations. The thick line in Fig. 8.44 demonstrates that the model fits the conductance data quite nicely.

Figure 8.45 shows the velocity–field relation obtained from this analysis (uppermost curve). The velocity–field relation shows pronounced resonances, which are labeled $i \rightarrow j(k)$, where the carriers tunnel from a state of the ith miniband to a state of the jth miniband spatially located k wells away. As can be intuitively expected, the $1 \rightarrow 2(1)$ resonance is the strongest and leads to the highest velocities.

The experimental data are compared with two different models. The circles show the result of a theoretical model which assumes that the carriers

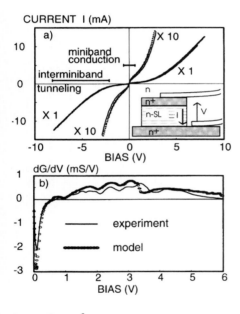

Fig. 8.44. *Top*: current as a function of the voltage at room temperature. *Bottom*: derivative of the conductance (d^2I/d^2V) vs. voltage in the experiment (*thin line*) and in the simulation (*thick line*) (from [70])

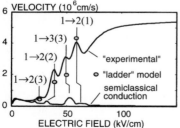

Fig. 8.45. Velocity–field relation used to model the I–V data. The *lower curve* assumes semiclassical conduction in the various minibands; the *open circles* mark peak velocities derived from the sequential ladder model (from [70])

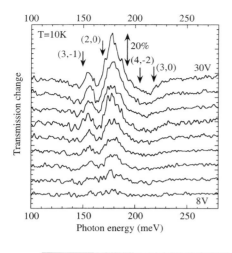

Fig. 8.46. Differential transmission spectra for increasing bias voltage. The peaks are rather independent of the field since they stem from the high-field domain, which becomes larger with increasing voltage, leading only to an increase in signal (from [71])

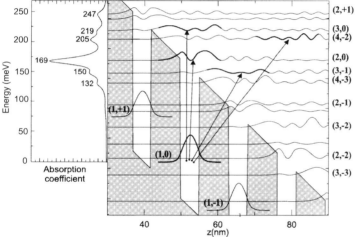

Fig. 8.47. Conduction band profiles, energy levels, and squared wave functions for a field expected in the high-field domain. The numbers on the *right* side represent a qualitative assignment of the transitions (*arrows*). The spectrum to the *left* is a model calculation of the absorption (from [71])

perform resonant tunneling processes to a higher miniband, followed by an interminiband relaxation process (a sequential ladder model). The lower solid line shows the result of a model which assumes semiclassical transport in the bands. At the high fields investigated here, this semiclassical transport associated with oblique transitions from well to well leads to much lower velocities than does the sequential ladder model.

In the infrared absorption experiments of Helm et al., a 50 Å/80 Å "shallow-well" GaAs/Al$_{0.29}$Ga$_{0.71}$As superlattice was employed. The sample shows I–V curves which correspond to a low-field and a high-field domain, which change their relative size with field. The infrared diffential transmission spectra displayed in Fig. 8.46 show pronounced resonances, which are due to vertical and oblique interminiband transitions. Figure 8.47 shows the energy structure of the sample and the various relevant transitions. Surprisingly, both in the transport measurements and in the optical spectra, a Zener resonance to the next-nearest-neighbor well state is very pronounced.

There are a number of further theoretical studies which are beyond the scope of our compilation. We therefore only mention here that the effect of Zener tunneling on transport has also been investigated theoretically by Rotvig et al. [73, 74] and by Zhao et al. [75–77].

8.7 Summary

In this chapter, we have reviewed electrical transport experiments, some of them with the inclusion of infrared optical excitation. A multitude of transport studies have clearly shown the key effect being negative differential velocity. Initially, it turned out that the actual experiments were often more difficult than expected, mainly owing to field inhomogeneities and/or domain formation. However, using various approaches, these hurdles have been overcome and many effects which are not easily accessible in bulk materials have been observed in the model system of a superlattice.

Experiments with infrared illumination have helped to clarify additional questions. Among those experiments are the observation of thermal saturation of band transport and the observation of the Shapiro effect, which corresponds to the inverse Bloch oscillator.

References

1. A. Sibille, in *Semiconductor Superlattices – Growth and Electronic Properties*, ed. H.T. Grahn (World Scientific, Singapore 1995), p. 29.
2. L. Esaki and R. Tsu, IBM J. Res. Dev. **14**, 61 (1970).
3. R.A. Suris and B.S. Shchamkhalova, Sov. Phys. Semicond. **18**, 738 (1984).
4. G. Brozak, M. Helm, F. DeRosa, C.H. Perry, M. Koza, R. Bhat, and S.J. Allen, Jr., Phys. Rev. Lett. **64**, 3163 (1990).
5. G. Brozak, C.H. Perry, M. Helm, F. DeRosa, M. Koza, R. Bhat, L.T. Florez, J.P. Harbison, and S.J. Allen, Jr., Proc. 20th Int. Conf. Phys. Semicond., eds. E.M. Anastassaki and J.D. Joannopoulos (World Scientific, Singapore 1991) p. 1617.
6. M. Artaki and K. Hess, Superlatt. Microstruct. **1**, 489 (1985).
7. A. Chomette and J.F. Palmier, Solid State Commun. **43**, 157 (1982).
8. I. Dharssi and P.N. Butcher, J. Phys.: Condens. Matter **2**, 4629 (1990).

9. J.F. Palmier, G. Etemadi, A. Sibille, M. Hadjazi, F. Mollot, and R. Planel, Surf. Sci. **267**, 574 (1992).
10. B. Deveaud, J. Shah, T.C. Damen, B. Lambert, and A. Regreny, Phys. Rev. Lett. **58**, 2582 (1987).
11. B. Deveaud, J. Shah, T.C. Damen, B. Lambert, A. Chomette, and A. Regreny, IEEE J. Quantum Electron. QE **24**, 1641 (1988).
12. B. Deveaud, A. Chomette, B. Lambert, A. Regreny, R. Romestain, and P. Edel, Solid State Commun. **57**, 885 (1986).
13. B. Lambert, B. Deveaud, A. Chomette, A. Regreny, and B. Sermage, Semicond. Sci. Technol. **4**, 513 (1989).
14. J. Feldmann, G. Peter, E.O. Göbel, K. Leo, H.J. Polland, K. Ploog, K. Fujiwara, and T. Nakayama, Appl. Phys. Lett. **51**, 226 (1987).
15. H.J. Polland, K. Leo, K. Ploog, J. Feldmann, G. Peter, E.O. Göbel, K. Fujiwara, and T. Nakayama, Solid State Electron. **31**, 341 (1988).
16. G.W. Oestel, Q.M. Onika, H. Found, Q.G. Not, S.T. File, and C. Unnersdorf, Phys. Rev. Lett. **67**, 1931 (1991).
17. H.J. Queisser, J. Appl. Phys. **37**, 2909 (1966).
18. J.F. Palmier, C. Minot, J.L. Lievin, F. Alexandre, J.C. Harmand, J. Dangla, C. Dubon-Chevallier, and D. Ankri, Appl. Phys. Lett. **49**, 1260 (1986).
19. L. Esaki and L.L. Chang, Phys. Rev. Lett. **33**, 495 (1974).
20. H.T. Grahn, Habilitationsschrift, Universität Erlangen and Max-Planck-Institut für Festkörperforschung Stuttgart (1992).
21. A. Sibille, J.F. Palmier, C. Minot, and F. Mollot, Appl. Phys. Lett. **54**, 165 (1989).
22. A. Sibille, J.F. Palmier, H. Wang, and F. Mollot, Phys. Rev. Lett. **64**, 52 (1989).
23. A. Sibille, J.F. Palmier, H. Wang, and F. Mollot, Appl. Phys. Lett. **56**, 256 (1990).
24. P. Gueret, Phys. Rev. Lett. **5**, 256 (1971).
25. P.S. Kop'ev, R.A. Suris, I.N. Uraltsev, and A.M. Vasiliev, Solid State Commun. **72**, 401 (1989).
26. R.A. Suris and B.S. Shchamkhalova, Sov. Phys. Semicond. **24**, 1023 (1990).
27. A. Sibille, J.F. Palmier, A. Celeste, J.C. Portal, and F. Mollot, Europhys. Lett. **13**, 279 (1990).
28. B. Mitra and K.P. Ghatak, Phys. Status Solidi B **164**, K13 (1991).
29. F. Aristone, A. Sibille, J.F. Palmier, D.K. Maude, J.C. Portal, and F. Mollot, Physica B **184**, 246 (1993).
30. H. Noguchi and H. Sakaki, Phys. Rev. B **45**, 12148 (1992).
31. B.W. Hakki, J. Appl. Phys. **38**, 808 (1967).
32. A. Sibille, J.F. Palmier, F. Mollot, H. Wang, and J.C. Esnault, Phys. Rev. B **39**, 6272 (1989).
33. A. Sibille, J.F. Palmier, H. Wang, and J.C. Esnault, Solid State Electron. **32**, 1455 (1989).
34. H. Schneider, K. von Klitzing, and K. Ploog, Europhys. Lett. **8**, 575 (1989).
35. H. Le Person, J.F. Palmier, C. Minot, J.C. Esnault, and F. Mollot, Surf. Sci. **228**, 441 (1990).
36. X.L. Lei, N.J.M. Horing, and H.L. Cui, Phys. Rev. Lett. **66**, 3277 (1991).
37. P. England, J.R. Hayes, E. Colas, and M. Helm, Phys. Rev. Lett. **63**, 1708 (1989).
38. R.R. Gerhardts, Phys. Rev. B **48**, 9178 (1993); R.R. Gerhardts, Solid State Electron. **37**, 681 (1994).

39. H.T. Grahn, K. von Klitzing, K. Ploog, and G.H. Döhler, Phys. Rev. B **43**, 12094 (1991).
40. A. Sibille, J.F. Palmier, M. Hadjazi, H. Wang, G. Etemadi, E. Dutisseuil, and F. Mollot, Superlatt. Microstruct. **13**, 247 (1993).
41. A. Chomette, B. Deveaud, A. Regreny, and G. Bastard, Phys. Rev. Lett. **57**, 1464 (1986).
42. F. Beltram, F. Capasso, D.L. Sivco, A.L. Hutchinson, S.N.G. Chu, and A.Y. Cho, Phys. Rev. Lett. **64**, 3167 (1990).
43. C. Rauch, G. Strasser, K. Unterrainer, E. Gornik, and B. Brill, Appl. Phys. Lett. **79**, 649 (1997).
44. C. Rauch, G. Strasser, K. Unterrainer, W. Boxleitner, E. Gornik, and A. Wacker, Phys. Rev. Lett. **81**, 3495 (1998).
45. R. Heer, J. Smoliner, G. Strasser, and E. Gornik, Appl. Phys. Lett. **73**, 3138 (1998).
46. V. Sankin and A. Naumov, Superlatt. Microstruct. **10**, 353 (1991).
47. V.I. Sankin and I.A. Stolichnov, JETP Lett. **59**, 744 (1994) [Pis'ma Zh. Eksp. Teor. Fiz. **59**, 703 (1994)].
48. V. Sankin, Superlatt. Microstruct. **18**, 309 (1995).
49. V. Sankin and I. Stolichnov, Superlatt. Microstruct. **23**, 999 (1998).
50. V.I. Sankin, Yu.A. Vodakov, and D.P. Litvin, Sov. Phys. Semicond. **18**, 1339 (1984) [Fiz. Tekh. Poluprovodn. **18**, 2146 (1984)].
51. R. Tsu and G. Döhler, Phys. Rev. B **12**, 680 (1975).
52. R. Tsu and L. Esaki, Phys. Rev. B **43**, 5204 (1991); Phys. Rev. B **44**, 3495 (1991) (erratum).
53. D. Miller and B. Laikhtman, Phys. Rev. B **50**, 18426 (1994).
54. S. Rott, N. Linder, and G.H. Döhler, Superlatt. Microstruct. **21**, 569 (1997).
55. A. Wacker and A.-P. Jauho, Phys. Scr. T **69**, 321 (1997).
56. A. Wacker and A.-P. Jauho, Phys. Rev. Lett. **80**, 369 (1998).
57. A. Wacker, A.-P. Jauho, S. Rott, A. Markus, P. Binder, and G.H. Döhler, Phys. Rev. Lett. **83**, 836 (1999).
58. S. Rott, P. Binder, N. Linder, and G.H. Döhler, Phys. Rev. B **59**, 7334 (1999).
59. S. Rott, N. Linder, and G.H. Döhler, Phys. Rev. B **65**, 195301 (2002).
60. A. Sibille, J.F. Palmier, and F. Mollot, Appl. Phys. Lett. **60**, 457 (1992); A. Sibille, J.F. Palmier, M. Hadjazi, and H. Wang, Proc. 21st Int. Conf. Phys. Semicond., eds. P. Jiang and H. Zheng (World Scientific, Singapore, 1992), p. 1190.
61. M. Helm, Semicond. Sci. Technol. **10**, 557 (1995).
62. K. Unterrainer, B.J. Keay, M.C. Wanke, S.J. Allen, Jr., D. Leonard, G. Medeiros-Ribeiro, U. Bhattacharya, and M.J. Rodwell, Phys. Rev. Lett. **76**, 2973 (1996).
63. A.A. Ignatov, K.F. Renk, and E.P. Dodin, Phys. Rev. Lett. **70**, 1996 (1993).
64. E.E. Mendez, F. Agullo-Rueda, and J.M. Hong, Phys. Rev. Lett. **60**, 2426 (1988).
65. P. Voisin, J. Bleuse, C. Bouche, S. Gaillard, C. Alibert, and A. Regreny, Phys. Rev. Lett. **61**, 1639 (1988).
66. A.A. Ignatov, E. Schomburg, K.F. Renk, W. Schatz, J.F. Palmier, and F. Mollot, Ann. Phys. **3**, 137 (1994).
67. E. Schomburg, A.A. Ignatov, J. Grenzer, K.F. Renk, D.G. Pavel'ev, Yu. Koschurinov, B.Ja. Melzer, S. Ivanov, S. Schaposchnikov, and P.S. Kop'ev, Appl. Phys. Lett. **68**, 1096 (1996).

68. S. Winnerl, E. Schomburg, J. Grenzer, H.-J. Regl, A.A. Ignatov, A.D. Semenov, K.F. Renk, D.G. Pavel'ev, Yu. Koschurinov, B. Melzer, V. Ustinov, S. Ivanov, S. Saposchnikov, and P. Kop'ev, Phys. Rev. B **56**, 10303 (1997).
69. S. Winnerl, W. Seiwerth, E. Schomburg, J. Grenzer, K.F. Renk, C.J.G.M. Langerak, A.F.G. van den Meer, D.G. Pavel'ev, Yu. Koschurinov, A.A. Ignatov, B. Melzer, V. Ustinov, S. Ivanov, S. Saposchnikov, and P.S. Kop'ev, Appl. Phys. Lett. **73**, 2983 (1998).
70. A. Sibille, J.F. Palmier, and F. Laruelle, Phys. Rev. Lett. **80**, 4506 (1998).
71. M. Helm, W. Hilber, G. Strasser, R. De Meester, F.M. Peeters, and A. Wacker, Phys. Rev. Lett. **82**, 3120 (1999).
72. C. Zener, Proc. R. Soc. London Ser. A **145**, 523 (1934).
73. J. Rotvig, A.-P. Jauho, and H. Smith, Phys. Rev. Lett. **74**, 1831 (1995).
74. J. Rotvig, A.-P. Jauho, and H. Smith, Phys. Rev. B **54**, 17691 (1996).
75. X.G. Zhao, G.A. Georgakis, and Q. Niu, Phys. Rev. B **54**, R5235 (1996).
76. X.G. Zhao, W.X. Yan, and D.W. Hone, Phys. Rev. B **57**, 9849 (1998).
77. X.G. Zhao and D.W. Hone, Phys. Rev. B **62**, 5010 (2000).

9 Bloch Oscillations in Other Systems

According to the description in Chap. 1, Bloch oscillations are a quite general phenomenon which is always present when there are wave states in a superposition of a periodic potential and a constant force (i.e. a linear potential). Besides semiconductor superlattices, three other systems have been investigated in recent years:

- Bloch oscillations and Zener tunneling have been investigated for atoms in optical lattices [1, 2].
- Bloch oscillations have been investigated in optical wave guide arrays [3,4].
- Bloch oscillations in a magnetic system have been proposed theoretically for the case of domain walls [5, 6].

In the following, we shall give a brief account of the first two sets of experimental observations and compare the main features with the experiments on semiconductor superlattices.

9.1 Bloch Oscillations of Atoms

The experiments on solid-state systems always address the dynamics of electronic wave packets. There is no reason why similar observations should not be possible if the dynamics are caused by the interference of larger entities of matter, for example atoms. Recently, the groups of Raizen and Salomon have demonstrated such dynamics in a series of experiments which are much more elegant and appealing than the experiments on the solid state.

The basic idea of the experiments is to place single atoms into light lattices, which are generated by the superposition of two counterpropagating light beams. Since the laser wavelength is typically in the visible range, the grating obviously has a periodicity in the hundreds of nanometers range, i.e. about an order of magnitude larger than in the typical solid-state experiments. The key difference, however, is the much weaker dephasing in the system: the weak coupling of the atoms to the outer world allows one to follow the dynamics over thousands of oscillations, which probably will never be possible in a solid state-experiment.

As an example, we describe here the atom Bloch oscillation experiment reported by Salomon's group [1, 7]. In this experiment, cesium atoms are placed in the potential minima of a laser-induced optical standing wave. The standing wave is generated by two diode lasers operating at about 850 nm, which is at the $6S_{1/2} \to 6P_{3/2}$ atomic transition. The potential depth of the standing wave is very small; it can be varied by the laser intensity between zero and about 5×10^{-11} eV! Therefore, the atoms need to be cooled to very low temperatures (in the μKelvin range) to achieve a defined initial energetic composition. This cooling is performed using the elaborate laser cooling techniques developed in atom spectroscopy over the last few decades.

The linear potential is then nonadiabatically turned on by the following method: one of the lasers which generate the standing wave is tuned in energy linearly with time. It is easy to realize that this leads to an acceleration of the grating towards the laser which has the decreasing frequency. This constant acceleration, in the frame of the standing wave, corresponds to a constant force along the grating. The laser experiments thus indeed perform the gedankenexperiment outlined in Chap. 1, where the electron is placed at $k = 0$ and the field is then turned on. The frequency of the Bloch oscillations was adjusted between 60 and 1900 Hz, i.e. about *ten orders of magnitude slower* than in the semiconductor experiments. The motion could be followed over thousands of oscillations, demonstrating the very weak dephasing in this experiment [1].

The atoms then start to perform Bloch oscillations, with a spatial range in the micron range (see Fig. 9.1 for a schematic picture of the experimental arrangement). The large spatial range is associated with the fact that, first, the periodicity of the potential is about one order of magnitude larger than in the solid-state experiments and, second, the much lower dephasing rate allows one to work at "fields" which are much lower than in those experiments.[1]

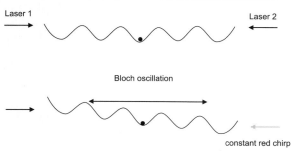

Fig. 9.1. Schematic illustration of the atom Bloch oscillations experiment

---

[1] In fact, it would be possible to visually observe the Bloch oscillations by tracing the position of the atoms as a function of time, for example by the scattering or fluorescence of the atoms. In the experiments described here, this has not been done, however.

In this system, the analysis of the wave packet motion is not performed by tracing the real-space motion, but by investigating the wave vectors of the atom wave packet as a function of delay time. For this k-space analysis, the periodic potential is turned off so that the atoms move freely. Then, they interact with a detuned Raman laser which transfers their linear momentum distribution into a distribution of the different fine-structure states. The relative occupation then yields the momenta of the atoms.

Figure 9.2 shows the distribution of the wave packet as a function of the delay time. It is obvious that the atom starts out as a wave packet that is rather well localized in k space. With increasing delay time, the wave packet increases its average k-vector linearly according to (1.4). A very intriguing observation can be made if the wave packet reaches the edge of the first Brillouin zone: while part of it leaves the Brillouin zone, another part of it enters the Brillouin zone, coming from negative k-vectors! This is simply caused by the fact that the band structure is invariant under translation by one reciprocal vector, and thus, a part which leaves a Brillouin zone can at the same time be seen to enter it.

The experiments reported in [1] have also beautifully shown the effect of bands with anharmonic dispersion. This effect can be investigated if the depth of the potential wells can be varied. If the height of the potential wells is rather large and the first band has a width which is small compared with the potential height, the band can be described well in the nearest-neighbor tight-binding description and is harmonic (see (1.6)). If the potential height is subsequently lowered, the band comes close to the upper edge of the potential and becomes anharmonic. Intuitively, the particles close to the upper edge are more "free" than if they are at the lower edge, i.e. the band curvature is

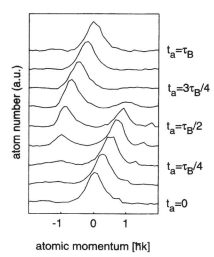

Fig. 9.2. Distribution of the atom in k-space as a function of the delay time (from [1])

larger since the effective mass must be lower. This leads to large velocities around the upper band edge and small velocities around the lower edge.

In the atom Bloch oscillation experiments, the potential height could be controlled in situ by the intensity of the light beams forming the standing wave potential. The anharmonicity was directly visible in the real-space velocity characteristics of the carriers (see Fig. 9.3): for deep potential wells (Fig. 9.3c), the velocities are low and the motion is harmonic; for the lowest wells (Fig. 9.3a), the velocity is much higher and clearly anharmonic.

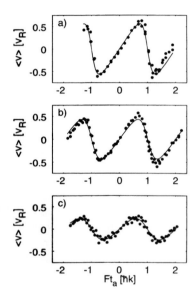

Fig. 9.3. Dynamics of the atom wave packet as a function of the potential depth, showing the effects of anharmonic band dispersion (from [1])

9.2 Bloch Oscillations in Optical Wave Guides

Another approach to investigating Bloch oscillations has been the study of a periodic array of waveguides [3, 4]. The particles which undergo Bloch oscillations are photons in this case. In a very simple picture, a single optical waveguide can be regarded as a well for the light, while the cladding is the barrier. The coupling of the guides in the array leads to a delocalization of the eigenstates. The linear potential can be realized in different ways: for instance, one can apply a linear temperature gradient, as done in the experiments described here [3], or one can vary the wave guide dimensions and/or spacing [4], or introduce a curvature [8]. In all cases, the delocalized states convert to an optical Stark ladder consisting of eigenstates.

The initial wave packet is generated by illuminating the front end of the array of waveguides arranged in parallel. Obviously, it is very easy to control

9.2 Bloch Oscillations in Optical Wave Guides 205

the initial spatial distribution of the wave packet. The Bloch oscillation is not performed in time, as in the solid-state and atom experiments, but in space: the light propagates down the wave guide array and performs a spatial oscillation perpendicular to the wave guide propagation direction.

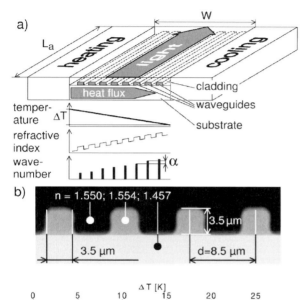

Fig. 9.4. Schematic illustration of the wave guide Bloch oscillation experiment (from [3])

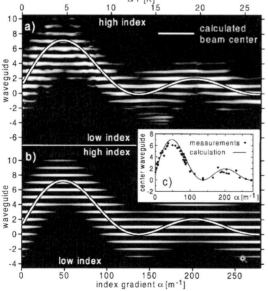

Fig. 9.5. Light traces in the wave guide Bloch oscillation experiment in the case of asymmetric excitation at the edge of the Brillouin zone. *Top*: experiment. *Bottom*: theory (from [3])

Figure 9.4 shows the setup of the experiment with temperature tuning [3]. The system consisted of 75 wave guides with a periodicity of 8.5 μm, a size of 3.5×3.5 μm^2, and a length of 4.5 cm.

For excitation of Bloch oscillations with a center-of-mass motion, one needs to have a wave packet which is quite well localized in $\boldsymbol{k}$-space. Therefore, one needs to excite an initial photon wave packet which is delocalized in real space. Figure 9.5 shows the result of such an experiment: here, the output light is plotted as a function of the wave guide number ($\boldsymbol{y}$ axis) and the temperature gradient ($\boldsymbol{x}$ axis).

By tuning the temperature gradient, one can tune the period (spatially, along the wave guide) of the Bloch oscillations. Thus, one can observe the spatial distribution of the Bloch wave packet as a function of its phase in $\boldsymbol{k}$-space. The data are thus roughly equivalent to following the spatial distribution as a function of time in the superlattice experiments. The data of Fig. 9.5 indeed show a very nice spatial oscillation. Note that the amplitude decrease to the right is not caused by some damping effect, but simply reflects the fact that the amplitude becomes smaller with increasing temperature gradient (i.e. field).

By excitation into one waveguide, one can simulate a breathing-mode experiment where one starts out with a wave packet which is localized in real

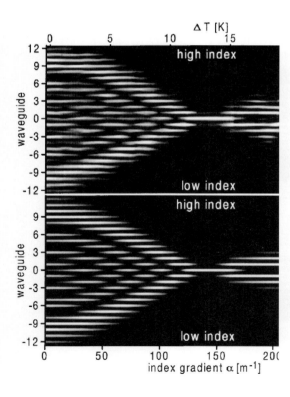

Fig. 9.6. Light traces in the wave guide Bloch oscillation experiment for the case of symmetric excitation in the Brillouin zone. *Top*: experiment. *Bottom*: theory (from [3])

space, i.e. distributed over all ***k***-space. Figure 9.6 clearly shows the breathing-mode behavior of the wave packet.

It is obvious that Bloch oscillation experiments in waveguides are a very direct approach. The system is comparatively easy to control, and the effects are directly visible. It can be expected that many more related phenomena, like Zener tunneling, will be investigated in such systems. Also interesting is the fact that one can study a rich new class of phenomena when one enters the high-intensity regime where nonlinear optical effects become important [4]. It should also be mentioned that the effects in waveguide arrays might have applications, for example for switching purposes.

References

1. M. Ben Dahan, E. Peik, J. Reichel, Y. Castin, and C. Salomon, Phys. Rev. Lett. **76**, 4508 (1996).
2. S.R. Wilkinson, C.F. Bharucha, K.W. Madison, Q. Niu, and M.G. Raizen, Phys. Rev. Lett. **76**, 4512 (1996).
3. T. Pertsch, P. Dannberg, W. Elflein, A. Bräuer, and F. Lederer, Phys. Rev. Lett. **83**, 4753 (1999).
4. R. Morandotti, U. Peschel, J.S. Aitchison, H.S. Eisenberg, and Y. Silberberg, Phys. Rev. Lett. **83**, 4756 (1999).
5. H.-B. Braun and D. Loss, J. Appl. Phys. **76**, 7177 (1994).
6. J. Kyriakidis and D. Loss, Phys. Rev. B **58**, 5568 (1998).
7. E. Peik, M. Ben Dahan, I. Bouchoule, Y. Castin, and C. Salomon, Phys. Rev. A **55**, 2989 (1997).
8. G. Lenz, I. Talanina, and C.M. de Sterke, Phys. Rev. Lett. **83**, 963 (1999).

10 Future Prospects and Possible Applications

In this chapter, we discuss possible device applications of Bloch oscillations and other high-field transport phenomena in semiconductor superlattices. We have to admit that this may seem to be a rather far-fetched discussion, since no such applications have been realized yet. Nevertheless, it seems to be worth discussing some basic issues in this field since, in the author's opinion, there is no doubt that there is a need for such devices, even if it is "only" a need for a tunable THz source for scientific spectroscopy. However, it remains to be seen whether there will be broad applications, for example Bloch emitters for high-speed free-space communications or for imaging in the THz range.

In the following, we shall first discuss some general issues involving the use of coherent electronic transport effects for device applications. Then, we shall specifically address the question of gain in semiconductor superlattices, which is a prerequisite for an electrically pumped source. Finally, we shall briefly address some possible device schemes.

10.1 Basic Considerations for Coherent Electronic Devices

In this discussion of basic considerations for devices, we follow here a recent description of coherent electronic effects in semiconductors [1]. Before we address the role of coherent electron transport in future devices, we need to briefly discuss the different forms of coherence which are relevant to possible devices. First, one has to distinguish between coherence in an electronic system and coherence in a photonic system:

- In the first form of coherence, the electronic wave functions are coherent, i.e. have a well-defined phase. For Bloch oscillations, this is the case when the ensemble is started with a well-defined phase relation, as is the case for short-pulse optical excitation.
- The second form of coherence is the case where the coherence is kept in the photon field, for example in a laser resonator. The coherent photon field can then be fed by an electronic system which is not in coherence.

In the following, we shall address both cases. First, we shall discuss the case of electronic coherence in devices, which is more challenging than the case of photonic coherence since the former is subject to the frequent scattering processes in a solid, whereas the latter is much more protected from perturbations.

Indeed, there are many restrictions on devices which use electronic coherence:

- Typically, device applications are limited to room temperature. This means that the electronic coherence times in semiconductor materials are restricted to the subpicosecond range owing to scattering by optical phonons.
- Up to now, it has only been possible to generate coherence in an ensemble (associated with a macroscopic polarization) by using coherent laser or THz sources, and not by electrical injection.

Owing to those difficulties, which are not known for conventional semiconductor devices, one should first consider whether coherent effects in superlattices allow applications which either are unique or offer parameters much better than those of present-day devices using other working principles. A brief (and definitely incomplete) list of such effects could be:

- Coherent effects allow the generation of radiation with tuning ranges which seem to be well beyond those of other sources. As we have discussed in the section about Bloch oscillations, tuning ranges of a factor of four have been shown already in the first experiments; later studies have achieved more than a factor of ten.
- Classical semiconductor devices make use of incoherent population effects. *Coherent polarizations* can be removed, for example by a subsequent laser pulse [2]. This might be used for ultrafast devices, where the avoidance of population buildup will lead to fast recovery times, since the comparatively slow removal by transport and/or recombination is not needed.
- A completely different area of application is quantum computing, where the information is stored in coherent quantum states. In principle, the wave packets created when Bloch oscillations are observed could function as quantum bits. However, the strong decoherence effects in semiconductors will make it difficult to achieve operation while keeping the coherence temporally and spatially over meaningful intervals. The topic of quantum computing is beyond the scope of this book. For a detailed discussion, see [3].

Although it is still quite speculative to talk about applications of coherent effects in superlattices, one should point out that applications based on semiconductors always have the advantage that they can be realized with a monolithic approach and profit from the sophisticated microstructurization techniques developed for classical electronics. For instance, if one is able to realize single qubits for a quantum computer on a semiconductor substrate, it seems as if it will be straightforward to integrate a large number of them on a chip and interface them to the outside classical-electronics environment.

10.2 Device Concepts Using Coherent Transport Effects in Superlattices

In the following, we shall address possible devices where coherent transport effects in superlattices are employed for detection and emission purposes. As discussed above, there are several areas where coherent transport and Bloch oscillations could be used in devices. We first discuss the application of nonlinear transport in superlattices, which has already been demonstrated to detectors. Then, we address possible emission devices.

10.2.1 Detector Applications

One possible detector principle is the current suppression due to dynamic localization: in a simple picture, the AC field "tilts" the superlattice miniband back and forth so fast that the carriers can move in neither direction. It was shown that this effect can lead to a nearly complete current suppression in superlattices [4, 5].

Subsequently, it was shown by Renk's group [6], in a series of experiments, that this nonlinear transport effect in superlattices can be used for the detection of far-infrared radiation and for frequency-mixing purposes.

Figure 10.1 shows the mesa sample arrangement (10.1 a), the current–voltage characteristics (10.1 b), and the signal of the detector in a video detection mode (10.1 c). The authors of [6] also performed correlation measurements which allowed them to determine the pulse duration of a free-electron laser, as visible in Fig. 10.2. The measured pulse durations were 11 and 5 ps, in the two experiments. The detector has a very high dynamic range, and a sensitivity which is comparable to that of intraband bulk semiconductor detectors [6]. The sensitivity of Schottky detectors is much higher, but their dynamic range is much lower than that of the superlattice detector.

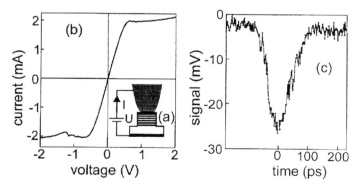

Fig. 10.1. (a) Mesa sample arrangement. (b) Current–voltage characteristics. (c) Signal of the detector in a video detection mode, displaying the FEL pulse (from [6])

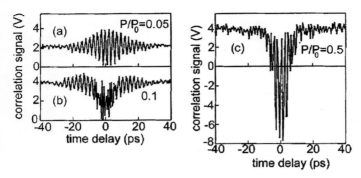

Fig. 10.2. Pulse of the free-electron laser, as measured with cross-correlations using the superlattice detector (from [6])

10.2.2 Free-Running Bloch Emitters

General considerations There are only a few sources of radiation in the range of about 1 to 10 THz, corresponding to photon energies of 4.2 to 42 meV or wavelengths from 300 to 30 μm. For lower frequencies, electronic devices are available; for higher frequencies, optical devices set in. The gap between the electronic and optical sources, the so-called "THz gap", could be closed by the Bloch oscillator.[1]

The use of Bloch oscillations as a source of tunable, coherent THz radiation would clearly be an impressive application of coherent effects in superlattices. The main advantage of a free-running Bloch emitter compared with other sources in this range would be the large possible tuning range and the compact size (provided that a monolithic realization is achieved).

Limiting factors for a Bloch emitter at room temperature are probably the coherence time, resulting in minimum photon energies of about 20 meV (at least for a free-running emitter), and the heterostructure offsets giving an upper limit of a few hundred meV, depending on the semiconductor materials system employed. The full tuning range determined by these restrictions will probably not be attainable in one single device. Nevertheless, tuning ranges of a factor of four have been shown, even in the first experiments that observed Bloch oscillations [7].

The free-running Bloch emitter as a source of tunable THz radiation has, in principle, been realized by the observation of tunable THz radiation emitted by an optically excited superlattice [8]. At first glance, however, such an optically pumped device is probably of little practical interest for several reasons:

- The efficiency of such a device is very low. Simply because of the fact that the optical excitation is in the interband range, large parts of the incoming

---

[1] One should mention here that quantum cascade lasers are rapidly moving into this gap, as discussed at the end of this chapter.

photon energy are usually lost, resulting in very low efficiency. Typically, the ratio between the emission photon energy and the pump photon energy, which is approximately given by eFd/E_g, is between 1:1000 and 1:100.
- Despite the rapid progress in ultrafast laser technology, femtosecond lasers are still rather bulky and expensive compared with those devices which are realized on a single chip.
- The emission of the device is rather broadband since, in contrast to a source based on stimulated emission, there is no resonator with a mode structure.

It should be noted that the broadband THz sources briefly described in Chap. 5 are also restricted in a similar manner when their efficiency and the necessity for an ultrafast laser source are considered.

However, there are a number of aspects which might overcome these limitations. The second limitation listed above will be less of a problem in the future owing to the advent of novel femtosecond laser sources. It is not very bold to predict that in the future, diode-pumped solid-state femtosecond lasers with the size of a a few cubic cm will be available at rather low cost. An even better solution would be the use of a femtosecond semiconductor laser, a device which has been demonstrated some time ago [9].

The broad emission spectrum cannot be avoided unless a feedback mechanism is employed, simply because the spontaneous-emission spectrum will generally be rather broad owing to the fast times for scattering by phonons at room temperature. The large width of the spectrum even at low temperatures is obvious from the data shown in Chap. 5.

To improve the efficiency, one has to look at the generation of radiation in more detail. The emission of the collectively excited transitions in the semiconductor can be understood as superradiance [10]. This effect is defined as the emission of many oscillators into the same optical mode. In contrast to superfluorescence, where the oscillators are out of phase, we consider here a coherent system where N oscillators are in phase. It can then be shown that the emission is enhanced by a factor N.

For a detailed understanding of the situation when a biased superlattice is optically excited, one has to perform a somewhat involved calculation of the superposition effects of a spatially distributed dipole ensemble, as done by Victor et al. [11, 12]. These authors treated the emission from a two-dimensional array of emitters uniformly distributed over a rectangle. It turns out that the emission rate is enhanced compared with a single oscillator, but by less than a factor N, even if it is assumed that the coherence is not lost in scattering events. This reduction of the superradiance intensity is a geometrical effect caused by the finite rectangular distribution. For typical parameters of optical excitation, it turned out that about 10^6 oscillators effectively participated in the superradiance. The radiative lifetime of the ensemble was then still about 1 ns, i.e. very long compared to the typical intraband dephasing times. Therefore, most of the energy stored in the coherently excited system is lost before it can be converted into THz radiation.

The efficiency of the oscillator is, accordingly, low: Victor et al. [11, 12] estimate that for an optical excitation in the mW power range which could lead to a possible THz emission of μW (owing to the ratio of the intraband to the interband energy of about 10^{-3}), the THz power actually emitted is only in the nW range. The overall efficiency is thus only about 10^{-6}. An absolute measurement of the emitted power,[2] which confirmed the theoretical estimates, was performed by Martini et al. [13].

How is it possible to increase the output power of an optically pumped THz emitter? The two most straightforward strategies to increase the power are not useful, since (i) much higher optical excitation powers lead to strong field screening and dephasing, and (ii) larger spot areas do not increase the power, since the phase relations needed for superradiance are not achieved. These results indicate that the simple approach of surface excitation of a superlattice sample is not feasible for realistic devices.

There are possible ways to overcome of these limitations, in part. The key idea behind these methods is a traveling-wave approach, i.e. the optical excitation of a particular dipole is performed right at the time when the THz wave passes it. One approach is an excitation beam with a tilted wavefront. As it turns out, the superradiant emission is also at its most efficient when the THz beam is emitted along the surface and the optical excitation wave front is tilted in such a way that the optical phase fronts hit the sample surface in phase with the THz wave [11, 12].

Another approach is the use of an external cavity which couples the THz radiation back to its spot of origin, thus using it several times to generate the superradiance [11, 12, 14–16]. A detailed discussion of this idea is contained in the PhD thesis of Martini [14]. One key point mentioned by these authors is the difference from a laser: in a laser, where inversion of the occupation probability is present, the coherence of the radiation is automatically guaranteed. In the superradiant amplification scheme, however, one needs to couple the radiation back with the correct phase, otherwise there is no amplification effect.

Using a feedback arrangement, Martini et al. [15] were able to prove the amplification of THz radiation. For their study, they used an InP THz surface emitter. However, the same principle should work for a Bloch oscillator as well.

In a first experiment, Martini et al. showed that the power emitted by an optically excited THz emitter is increased by nearly a factor of three if it is

---

[2] These data on the absolute emitted power were then compared with theory to determine the absolute amplitude of the Bloch oscillations. It turns out that the theoretical calculation of the emitted power is quite involved, and effects such as the size and shape of the laser excitation and absorption of the radiation in the contact layers need to be carefully included to describe the emitted radiation. Additionally, the quantitative measurement of the emitted power is also rather involved. If one takes the many uncertainties and approximations of this analysis into account, one might speculate about the relevance of the amplitude data obtained in this manner.

irradiated with THz background radiation from another emitter; this is close to the results expected from superradiance theory.

In a second experiment, these authors demonstrated a cavity geometry (Fig. 10.3). Here, the signal from a surface emitter generated by femtosecond laser pulses is fed back by a ring resonator. The cavity length is adjusted suitably to match the repetition rate of the excitation laser, thus ensuring that the amplification field arrives with the correct phase.

Figure 10.4 a shows the THz field as a function of the delay time which was measured by coupling out a small part of the field from the cavity by means of a semitransparent mirror. If the cavity feedback is turned on, the THz electric field is increased by more than a factor of two, corresponding to more than a factor of six in the THz power. This is close to the theoretically expected result, which is limited mainly by the quite high loss of the resonator of about 50% per round trip.

Figure 10.4 b shows the field power as a function of frequency with and without feedback. It is obvious that the amplification is rather broadband, with a maximum around 600 GHz. These results show quite nicely that some of the limits of the optically pumped Bloch oscillator can be overcome by a clever arrangement to extract the power using superradiance.

Electrically Pumped Emitters Most of the problems associated with an optically pumped Bloch emitter would be solved if electrical pumping could be applied. Superlattices have been used to demonstrate electrically pumped GHz sources [17, 18]. However, these sources are based on the motion of domains through the superlattice structure. Although they depend indirectly on Bloch oscillations, since they use the nonlinearity of the current–voltage curve, these oscillators operate in the tens to hundreds of GHz range, well below the typical Bloch oscillation frequencies in the THz range that we consider here. Therefore, we shall not discuss these sources in more detail. However, it should be mentioned that model calculations which show that the frequency of the current oscillations could be raised up to 500 GHz by quenching of the domains have recently been published [19].

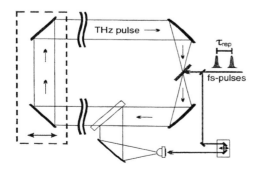

Fig. 10.3. Ring resonator setup with extracavity detection. The round-trip time in the cavity can be adjusted by moving two mirrors to match the repetition rate of the excitation laser (from [15])

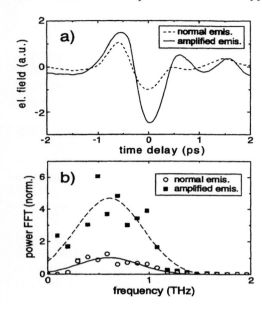

Fig. 10.4. (**a**): Detected electric-field amplitude from the ring resonator without (*dashed line*) and with (*solid line*) resonator feedback. (**b**) Power Fourier transform for emission without (*open circles*) and with (*filled squares*) resonator feedback (from [15])

When one considers an electrically pumped Bloch emitter, it seems that there are basically two schemes by which such an operation can be achieved:

- In one approach, electronic wave packets are injected in such a way that their composition is similar to the optically generated wave packets. The electronically injected wave packets should then perform Bloch oscillations and spontaneously emit radiation as in the case of optical injection. This free-running source would emit tunable THz radiation. If the injected electronic wave packets have random phase, the device will produce incoherent radiation; if there is a mechanism to phase-lock the wave packets, the device could produce coherent radiation as well.
- In the second, conceptually much simpler approach, a system which can perform Bloch oscillations is put in a cavity which couples the radiation back. If the Bloch-oscillating system produces gain, coherent radiation will be emitted. This "Bloch laser" would have the advantage of potentially narrow line widths and would emit coherent THz radiation.

The first operation principle requires that the electrical injection mechanism creates electronic wave packets which are temporally short enough that their energy spread is larger than the Wannier–Stark ladder splitting. This would require a mechanism that was able to create sub-picosecond electronic wave packets. It seems unlikely that such an effect could be easily found using nonlinear transport effects such as, e.g., the Gunn effect: the typical timescale of the Gunn effect is the return time from the indirect valleys, which is on the order of a few ps in GaAs [20], for example.

Oscillators based on resonant-tunneling diodes have reached operating speeds which come close to the requirements discussed above [21, 22]. However, a device which uses this injection mechanism for the wave packets is still limited in that the emission will be broadband and that the device can operate only in pulsed mode. In the elegant electrical-injection experiment by the Gornik group [23–25] discussed in Sect. 8.4.4, the spectral width of the electron beam which was injected into the base was on the order of 20 meV. However, this width was caused by inhomogeneous broadening of the injection, not by the intrinsic homogeneous width of the wave packets injected into the superlattice.

However, it seems that it is not necessary to generate wave packets by mechanisms similar to those in the optical experiments. It was first shown theoretically by Kazarinov and Suris [26, 27] that an electrically pumped superlattice could emit far-infrared radiation. This was indeed observed by Helm et al. [28]. However, the intensity of the radiation was rather weak.

The key question for the second kind of device, the Bloch laser, is whether there is gain in a superlattice and whether it is sufficient to overcome the losses which will occur in a device structure. These questions are addressed in the next section.

10.3 Bloch Lasers, or Is There Gain in a Biased Superlattice?

The ideal approach for realizing an electrically pumped Bloch emitter would be based on gain and would implement some scheme which couples the radiation back into the superlattice. Such a scheme has been proposed for coupled quantum wells in a symbolic manner by Luryi [29], where the device contacts were connected by a wire. In practice, the feedback mechanism could be realized by placing the superlattice in a cavity.

The prerequisite for getting such a device to operate as a laser is the presence of gain at the operating wave length. It is known that devices which have negative differential conductance

$$G(\omega) = \text{Re}\{dI/dV\} < 0 \tag{10.1}$$

at a given frequency ω are able to provide gain at that frequency. It is therefore to be expected that superlattices where negative differential conductivity has been experimentally proven (see Chap. 8) should provide gain and therefore be potential gain elements for a THz laser. However, up to now, it has not been shown experimentally that gain exists in an electrically pumped superlattice device at the Bloch frequency.

The use of gain in a Bloch laser is closely connected to the question of whether a device based on gain in a superlattice can operate in a stable regime. The gain in the superlattice is equivalent to a negative real part of the complex dynamic conductivity $\sigma(\omega)$. However, it has been shown that a homogeneous electric-field distribution is unstable when this negative AC conductivity extends down to zero field (see, e.g., [30]). This result is actually intuitively clear: if a homogeneous ensemble flows with constant current across the device, any perturbation which, for instance, locally reduces the field will cause an increase of the perturbation. Thus, domains will form, as is the case for the well-known Gunn effect. A lucid discussion of such instability effects for the Gunn effect and for an avalanche-induced process has been given by Kroemer [31].

From these considerations, it is thus necessary to have a system which has both gain at the operating frequency (i.e. near the Bloch frequency) and positive conductivity at lower frequencies, thus allowing stable operation. In the next two subsections, we shall address these questions.

10.3.1 Existence of Gain

A number of theoretical studies have addressed the question of whether there is gain in superlattices. As discussed in Chap. 8, the existence of negative differential conductivity was established by Esaki and Tsu [32]. As the next step, Ktitorov et al. [33] studied the complex conductivity $\sigma(\omega)$ of a biased semiconductor superlattice. Their results showed that the real part of the complex conductivity $\text{Re}\{\sigma\}$ which is already negative at $\omega = 0$, becomes increasingly negative with increasing frequency, reaches a minimum near the Bloch frequency, and then changes sign at the Bloch frequency. The most promising choice of the operating frequency of a Bloch laser would be at this minimum, somewhat below the Bloch frequency.

The dynamic conductivity of a superlattice as a function of frequency ω and electric bias field F can be written as

$$\sigma(\omega, F) = \frac{\sigma_0}{1 + \omega_B^2 \tau^2} \times \frac{1 - i\omega\tau - \omega_B^2 \tau^2}{\omega_B^2 \tau^2 + (1 - i\omega\tau)^2} \times \frac{I_1(\Delta/2k_B T)}{I_0(\Delta/2k_B T)}, \quad (10.2)$$

where τ is the scattering time, $\omega_B = eFd/\hbar$ is the Bloch frequency, Δ is the miniband width, and I_j is a modified Bessel function.

Figure 10.5 shows the dynamic conductivity $\sigma(\omega, F)$ as a function of $\omega\tau$ for various products of the Bloch oscillation frequency and the scattering time, $\omega_B \tau$. For $\omega_B \tau = 0$, i.e. in the low-field limit $F \to 0$, the dynamic conductivity starts at a positive value and then decays with increasing frequency. This is the standard Drude behavior. The behavior changes completely with increasing $\omega_B \tau$: for $\omega_B \tau = 1$, the dynamic conductivity drops to zero at $\omega = 0$; for $\omega_B \tau > 1$, there is a region of negative dynamic conductivity, i.e. there is gain.

It thus follows that the condition $\omega_B \tau > 1$ plays a crucial role in the development of gain. This can be understood as follows: for $\omega_B \tau < 1$, Bloch oscillations are suppressed since the carriers are scattered before a Bloch

10.3 Bloch Lasers, or Is There Gain in a Biased Superlattice?

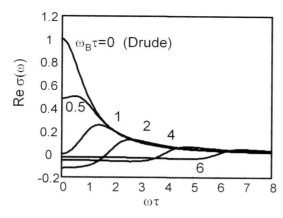

Fig. 10.5. Real part of the dynamic conductivity of a biased superlattice as a function of the frequency of the electric field, given in units of $\omega\tau$. The product of the Bloch oscillation frequency and relaxation time, $\omega_B\tau$, is varied from 0 (Drude limit) to 6

oscillation develops; for $\omega_B\tau > 1$, one reaches the quantum transport regime and Bloch oscillations take place. In other words, for smaller products $\omega_B\tau$, the optical density of states is a continuum; for $\omega_B\tau$ above 1, the Wannier–Stark resonances have developed and quantum transitions between them are possible.

Figure 10.6 shows the real part of the dynamic conductivity in the quantum regime ($\omega_B\tau = 10$) as a function of the frequency ω, given relative to the Bloch frequency ω_B. There is a negative dynamic conductivity, i.e. gain, for frequencies from zero to the Bloch oscillation frequency. The maximum of the gain is achieved close below the Bloch oscillation frequency ω_B. In a qualitative consideration based on the Wannier–Stark ladder picture, this behavior can be easily understood as follows. For photon energies below the Wannier–Stark ladder spacing (i.e. $\omega < \omega_B$), transitions down the Wannier–Stark ladder are possible since the missing energy to match a full step in the Wannier–Stark ladder can be provided by a scattering process. In contrast, transitions up the Wannier–Stark ladder are less likely, since the missing energy needs to supplied by phonon absorption. This scenario has already been outlined in Fig. 8.43.

Since the early studies of Esaki and Tsu [32] and Ktitorov et al. [33], many further theoretical studies have been performed. A theoretical study by Bastard and Ferreira [35] addressed possible gain in superlattices. In particular, it was predicted that there should be no gain from a biased superlattice at the Bloch frequency. The key point behind the argument was the balance of absorption and emission: if the Wannier–Stark ladder is populated with carriers, transitions one step "down" the ladder are as equally probable as transitions one step "up" the ladder, so that absorption and stimulated emission will cancel. This follows from the translation symmetry of the system and actually is consistent with the results for the dynamic conductivity discussed above: at the Bloch frequency ω_B, the results of Ktitorov et al. [33] and (10.2) predict zero dynamic conductivity. As discussed above, this symmetry

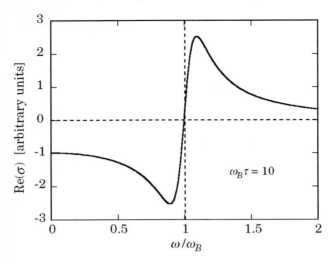

Fig. 10.6. Real part of the dynamic conductivity of a biased superlattice as a function of the electric field. The product of the Bloch oscillation frequency and the relaxation time $\omega\tau$ has been set equal to 10 (from [34])

is broken for frequencies below the Bloch frequency because of the relaxation processes.

The result of Bastard and Ferreira is seemingly in contradiction to the observation of THz radiation at the Bloch frequency from an optically pumped superlattice [8]. However, as discussed above, the emission in the optically pumped case can be regarded as spontaneous emission or superfluorescence, and thus is not in contradiction to the zero gain at the Bloch frequency.

Ignatov et al. [36] have investigated the behavior of a superlattice under both a DC and an AC field, i.e. for a field

$$F(t) = F_0 + F_\omega(\omega t). \tag{10.3}$$

These authors calculated an "oscillator efficiency", defined as the ratio of the power generated in the alternating field P_{alt} divided by the power dissipated in the DC bias field P_{bias}

$$\eta(\omega) = \frac{P_{\text{alt}}}{P_{\text{bias}}} = \frac{\int V(t) E_\omega \cos \omega t \, dt}{\int V(t) E_0 \, dt}, \tag{10.4}$$

where V is the average ensemble velocity. If $\eta > 0$, the superlattice absorbs energy; if $\eta < 0$, the AC field gains energy from the static field and the system displays gain.

Figure 10.7 shows the oscillator efficiency η for a superlattice as a function of the frequency for various electric fields, given in units of the critical field. The gain region extends from zero frequency up to the Bloch frequency (denoted by f_Ω in the figure); above, there is strong loss for all fields. For higher fields, there are resonances in the efficiency at subharmonics of the Bloch frequency.

Recently, Willenberg et al. [38] have investigated in a theoretical study the gain for both quantum cascade structures and superlattices, using a density

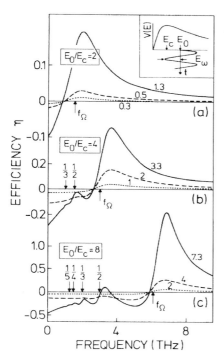

Fig. 10.7. Superlattice oscillator efficiency for different relative fields, given in units of the critical field F_C defined in (8.6). Note that in the figure, the critical field is denoted as E_C (from [36])

matrix formalism. The calculations corroborate the predictions of the semiclassical miniband picture, according to which gain is predicted for photon energies below the Bloch oscillation frequency.

Experimentally, information about gain in superlattices is very scarce. As we have discussed in Sect. 4.4, the linear coherent transport of the carriers observed in the experiments on the self-induced Shapiro effect by Löser et al. [37] proves the existence of gain at the Bloch frequency. However, the experiments with pulsed optical excitation address a nonequilibrium situation, and the gain is only transient. Therefore, these effects will most likely not be useful for devices.

The only experimental study which has directly addressed the gain of a superlattice under a steady-state electrical current was performed by Allen et al. [39]. In this experiment, the authors used a free-electron laser to illuminate a suitably designed superlattice sample. Figure 10.8 shows the setup of the experiment. To achieve efficient amplification, the authors used a grid of superlattices which were connected in series, forming a quasi-optical array of the kind known in the field of microwave antennas. The resonator was formed by an inductive grid on the back surface of the sample and by a metal reflector placed on top of the sample. This reflector could be moved up and down to achieve resonance with the free-electron laser photon energy.

222   10 Future Prospects and Possible Application

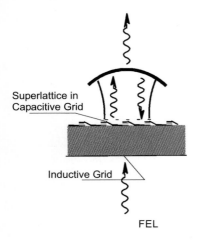

Fig. 10.8. Schematic illustration of a superlattice grid sample with a resonator. The resonator is formed by an inductive grid at the bottom of the substrate and a metal reflector on top of the sample (from [39])

The authors of [39] then measured the change in the transmission of the free-electron laser radiation when a bias voltage was applied to the superlattice sample. Figure 10.9 shows transmission changes measured at frequencies of 325 GHz and 2.55 THz. For the lower frequency (left panel), the transmission change is positive and rather large. However, it is still much smaller than the result expected from theory assuming a homogeneous field. If the authors assumed a field instability leading to the formation of a high-field and a low-field domain, they obtained a theoretical prediction closer to experiment. The remaining difference could possibly be explained by parasitic effects [39].

For the higher frequency of 2.55 THz, the data in Fig. 10.9 show a negative transmission change, which is much smaller than the positive change at the

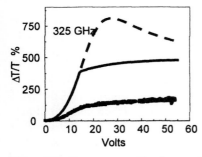
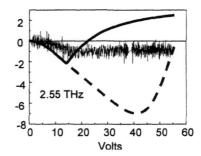

Fig. 10.9. Experimentally observed transmission changes (*solid lines* with noise) as a function of bias voltage across the superlattice sample for frequencies of 325 GHz (*left panel*) and 2.55 THz (*right panel*). The *dashed lines* are theoretically modeled results assuming a homogeneous electric field across the superlattices; the *smooth solid lines* are theory assuming domain formation (from [39])

lower frequency. For the theory without domains, the expected signal is much larger. If domains are included, the expected signal looks qualitatively very different and crosses zero at small voltages. Despite the very sophisticated approach and the clever design of the setup, the experiments thus do not give a fully conclusive picture. The main reason for that is probably the formation of domains, which leads to undefined fields in the device and thus a superposition of contributions. Also, the data presented show a reduction of absorption, but no proof of gain.

The problem of domain formation not only hinders investigative experiments, but also is the main problem for the operation of a device, as we shall discuss in the following. As already pointed out above, the stable operation of a device is only possible if the real part of the dynamic conductivity becomes zero or positive for low frequencies. In the following, we discuss proposals for stable operation of a Bloch oscillator.

10.3.2 Stable Operation of a Bloch Oscillator

In recent years, there have been three different proposals to overcome the instability problem of the Bloch oscillator:

- Wacker et al. [40] have shown that in weakly coupled superlattices, the conditions for stable operation seem to be intrinsically fulfilled.
- It was suggested by Kroemer [34] that a sufficiently strong AC excitation of the superlattice could lead to a stable operating regime.
- Finally, it was suggested by Allen et al. [39] that selective contacts to the superlattice layers might allow one to damp the instabilities.

The work of Wacker et al. [40] investigates theoretically the dynamic conductivity of weakly doped and weakly coupled superlattices. By using a sequential-tunneling model, these authors show that the dynamic conductivity, which is negative slightly below the Bloch oscillation frequency, becomes positive in these samples at low frequencies.

The authors of [40] describe the current response by the equation

$$I(t) = I_0 + \sum_{h=1}^{\infty} \left[I_h^{\cos} \cos(h\omega t) - I_h^{\sin} \sin(h\omega t) \right] . \tag{10.5}$$

Figure 10.10 shows various results of the theory. The key point is displayed in Fig. 10.10 b which shows that for the given sample, the in-phase alternating-current prefactor $I_1^{\cos}$ (which describes the real part of the dynamic conductance) is negative over a large frequency range, but positive at $\omega = 0$. Therefore, it is possible that the system can deliver gain at the Bloch oscillation frequency ($\hbar\omega_B = 8$ meV in Fig. 10.10 b) while it is stable against domain formation. Figure 10.10 c,d show the same current factor as a function of the DC and AC fields. It is obvious that the gain is observed over comparatively large field regions.

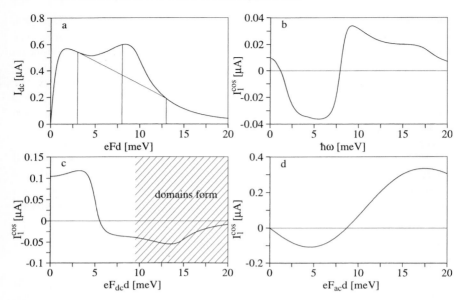

Fig. 10.10. (a) Current–field curve without irradiation. (b) Real part of the dynamic current response $I_1^{\cos}$ for a fixed static-field energy of $F_{\mathrm{dc}}d = 8$ meV and an oscillating-field energy $F_{\mathrm{ac}}d = 1$ meV. (c) As (b), but for fixed $F_{\mathrm{ac}}d = 1$ meV and $\hbar\omega = 5$ meV. (d) As (b) for $F_{\mathrm{dc}}d = 8$ meV and $\hbar\omega = 5$ meV (from [40])

The reason why this type of behavior could provide conditions for stable operation is the plateau region shown in Fig. 10.10 a, where the DC current increases a second time. This plateau was also observed experimentally and has been attributed to an impurity band. It allows one to choose the operating point in a positive differential regime while a derivative over a finite distance is negative.

The proposal of Kroemer [34] uses a dynamic scheme for the suppression of domain buildup. In this proposal, a strong additional AC field modulates the overall field in such a way that during each field cycle, the field drops low enough to reach a field region where the sample has the ordinary positive conductivity. During this time, the buildup of domains is damped strongly enough to prevent the instability altogether. A similar approach has been used previously for the Gunn effect and is referred to as the LSA-mode (limited space charge accumulation mode). The basic idea of this is illustrated in Fig. 10.11, which shows the operating mode between fields above and below the critical field. The overall field across the sample is expressed as

$$F = F_0 + F_\omega \cos(\omega t). \tag{10.6}$$

The theory then calculates the current–voltage curves, where a series expansion in terms of the number of photons n is employed. In the following figures, the results for the first term $n = 1$ and for all terms are displayed.

10.3 Bloch Lasers, or Is There Gain in a Biased Superlattice?

Figure 10.12 shows the calculated DC current as a function of the static field F_0, normalized to an equivalent field F^* which is defined by the field drop over one superlattice period corresponding to the photon energy at the operating frequency ω; i.e.

$$F^* = \hbar\omega/ed. \tag{10.7}$$

Since the THz photon energy needs to be smaller than the Wannier–Stark ladder splitting, one can see that the operating frequency of the Bloch oscillator must be at least slightly above F^*.

According to Fig. 10.12 the slope of the DC conductivity is positive around F_0/F^*. In other words, the DC conductivity, which starts to show a region of negative differential conductivity even at rather low field (since $F^*/F_C = 10$, this region starts at $F_0/F^* = 0.1$), has a second region of positive differential conductivity at a much higher field. This is caused by the presence of the AC field, which suppresses the domain formation. The results thus nicely show that stable operation of a device should be possible. Since the negative dynamic conductivity at the operation point of the Bloch oscillator requires $F_0 > F_C$, the condition that the DC conductivity is positive and the AC conductivity at the operation frequency ω is negative is satisfied for a static bias field F_0 in the range

$$F^* < F_0 < F^* + F_C. \tag{10.8}$$

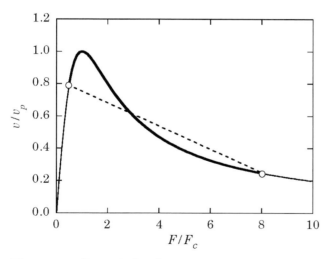

Fig. 10.11. Static drift velocity vs. field characteristics in normalized units. The *thick part* of the line illustrates the limited space charge accumulation operating mode; the current and field oscillate along this line. The *dashed line* illustrates a load line with negative slope which leads to power delivery to the external circuit (from [34])

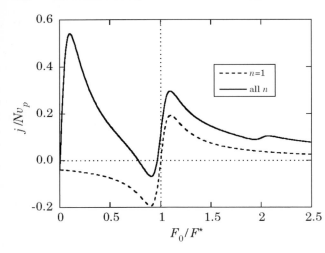

Fig. 10.12. DC current obtained by weighted summation over the photon replicas of $v(F)$, assuming the default parameters $F^*/F_C = 10$ and $F_\omega/F^* = 1$. The *solid line* is the main result, where all terms of a series expansion are included (from [34])

The requirement to operate in this field regime is rather obvious if one once more consults Fig. 10.11: the overall field during an oscillation must fall below the critical field, i.e. $F_0 - F_\omega < F_C$. With the parameters chosen here ($F_\omega/F^* = 1$), this is exactly the case for the field range given in (10.8). The suitability of the approach proposed by Kroemer is also visible in the conductivity, which shows a broad range of positive DC conductivity around the operating point [34].

Figure 10.13 shows the AC conductivity at the operating frequency ω as a function of the static bias field. It is obvious that the AC conductivity is negative for $F_0 > F^*$, i.e. there is AC gain in the region where the DC conductivity is positive. Despite the somewhat involved principle of the operation, it seems obvious that domain suppression by an additional AC field works, at least in theory. In a real device, the strong AC field might actually be provided by the THz laser field itself: if the fluctuations are large enough, the device might be self-starting.

Another approach to suppressing domains during the operation of a Bloch oscillator has been given by Allen et al. [39], who proposed to shunt the layers of the superlattice with a resistor which has a resistance low enough to make the DC conductivity positive, while keeping a negative AC conductance at the operating point below the Bloch oscillation frequency. However, such an approach might be difficult because it requires a very sophisticated sample growth and processing procedure to achieve the contacts to each layer.

To summarize this section, we have to conclude that there is still an insufficient understanding of the experimental and theoretical aspects of gain in biased superlattices. There is other theoretical work, for instance a study by Lei et al. [41] which predicts quite subtle AC effects on transport in superlattices which have not yet been considered in the models described above. Clearly, there is more work needed to clarify these important issues.

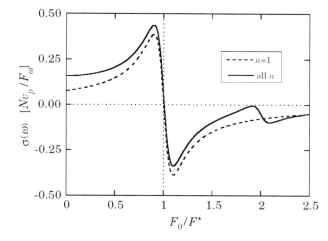

Fig. 10.13. Normalized AC conductivity as a function of the static bias field, expressed as F_0/F^* (from [34])

10.4 Comparison with Other Potential Sources of THz Radiation

When discussing the realization of a Bloch oscillator, one also has to take into account that such a device will have, at least in part of the frequency band it could reach, strong competition from other sources of THz radiation. Recently, semiconductor heterostructures have been used to realize lasers which involve only transitions in the conduction band of semiconductors. These structures have achieved room-temperature operation with quite high powers and have quickly surpassed lead-salt lasers which were previously the only semiconductor laser sources in this frequency regime.

The initial quantum cascade structures [42] used laser transitions between coupled quantum wells. More recently, a superlattice laser has been realized [43]. This laser uses the interband transition between the lower edge of the second miniband and the upper edge of the first miniband, both located at the edge of the mini-Brillouin zone. This scheme has the advantage of easily achieving inversion, owing to the fact that the intraband relaxation time (which populates the upper laser level and depopulates the lower laser level) is fast compared with the interminiband relaxation time. The quantum cascade devices reported have reached room-temperature operation and very high power, in the watt range.

Recently, quantum cascade lasers were successfully scaled into the THz frequency region. Following earlier theoretical predictions about the scalability of a device using interminiband transitions [44], Köhler et al. [45, 46] recently achieved lasing operation at 4.4 THz.

Figure 10.14 displays the energy band structure of the device. The squared wave functions (probability densities) of the electronic transitions in the many-well devices are also plotted. In the active superlattice region, the laser

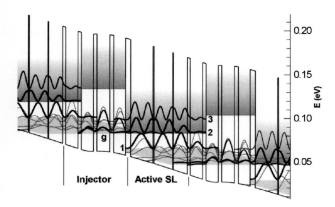

Fig. 10.14. Band energy scheme and probability densities of the interminiband laser design (from [44])

transition is between states in the second miniband (labeled 2) to states in the first miniband (labeled 1), with an interminiband spacing of 18 meV. The optical transition is vertical and has a rather large dipole moment of 7.2 nm. However, besides the band gap engineering, the design of the THz laser requires the solution of quite involved problems to achieve efficient waveguides [45].

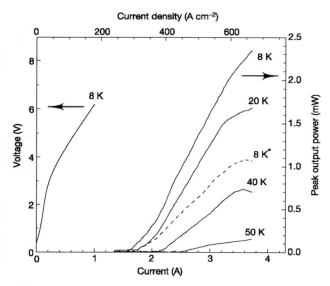

Fig. 10.15. Current–light and current–voltage curves of a 180 μm wide, 3.1 mm long laser structure under pulsed operation (from [45])

Figure 10.15 displays the current–voltage curve (left part) and the light–current curve (right part) of a laser structure. The data were taken in pulsed operation, applying 100 ns pulses with a repetition rate of 333 Hz. The threshold current density is quite low (290 A/cm^2). The device achieves an output of a few milliwatts at 8 K and lases up to 50 K. Similar data have recently also been obtained by Rochat et al. [47].

It is clear that any Bloch oscillator device has to compete with these very efficient intraband lasers. The only key advantage that a Bloch oscillator or laser could have over those structures is a large insitu tunability of the emission frequency. However, it remains to be seen whether the Bloch oscillator, if it is ever realized, will be a successful competitor against quantum cascade designs.

10.5 Summary

In this chapter, we have discussed possible applications of semiconductor superlattices. Despite their enormous importance for the understanding of the basic transport properties in periodic potentials, there have been no direct applications up to now. However, there are possibilities for use as infrared sensors.

The most exciting application, the tunable Bloch laser, is still a subject of uncertainty. While there have been many predictions that a biased superlattice can provide gain, the problems associated with the intrinsic instability of the current flow in a biased superlattice have not been solved experimentally. Recently, there have been some interesting proposals which might help to overcome this problem. It can be easily predicted that this topic will remain a lively area of research for some time to come.

References

1. K. Leo, "Coherent Effects in Semiconductor Heterostructures", in *Handbook of Advanced Electronic and Photonic Materials and Devices*, ed. H.S. Nalwa (Academic Press, San Diego 2001), p. 279.
2. A.P. Heberle, J.J. Baumberg, and K. Köhler, Phys. Rev. Lett. **75**, 2598 (1995).
3. A. Steane, Rep. Prog. Phys. **61**, 117 (1998).
4. A.A. Ignatov, E. Schomburg, K.F. Renk, W. Schatz, J.F. Palmier, and F. Mollot, Ann. Phys. **3**, 137 (1994).
5. S. Winnerl, E. Schomburg, J. Grenzer, H.-J. Regl, A.A. Ignatov, A.D. Semenov, K.F. Renk, D.G. Pavel'ev, Yu. Koschurinov, B. Melzer, V. Ustinov, S. Ivanov, S. Saposchnikov, and P. Kop'ev, Phys. Rev. B **56**, 10303 (1997).
6. S. Winnerl, W. Seiwerth, E. Schomburg, J. Grenzer, K.F. Renk, C.J.G.M. Langerak, A.F.G. van den Meer, D.G. Pavel'ev, Yu. Koschurinov, A.A. Ignatov, B. Melzer, V. Ustinov, S. Ivanov, S. Saposchnikov, and P.S. Kop'ev, Appl. Phys. Lett. **73**, 2983 (1998).

7. K. Leo, P. Haring Bolivar, F. Brüggemann, R. Schwedler, and K. Köhler, Solid State Commun. **84**, 943 (1992).
8. C. Waschke, H.G. Roskos, R. Schwedler, K. Leo, H. Kurz, and K. Köhler, Phys. Rev. Lett. **70**, 3319 (1993).
9. Y.K. Chen, M.C. Wu, T. Tanbun-Ek, R.A. Logan, and M.A. Chin, Appl. Phys. Lett. **58**, 12 (1991).
10. R.H. Dicke, Phys. Rev. **93**, 99 (1953).
11. K. Victor, H.G. Roskos, and C. Waschke, J. Opt. Soc. Am. B **11**, 2470 (1994).
12. H.G. Roskos, Habilitationsschrift, RWTH Aachen (1995).
13. R. Martini, G. Klose, H.G. Roskos, H. Kurz, H.T. Grahn, and R. Hey, Phys. Rev. B **54**, R14325 (1996).
14. R. Martini, Dissertation, RWTH Aachen, 1999, Berichte aus der Physik D82 (Shaker Verlag, Aachen 1999).
15. R. Martini, F. Hilbk-Kortenbruck, P. Haring Bolivar, and H. Kurz, "Inversionless Amplification of Coherent THz Radiation", Proc. IEEE 6th Int. Conf. Terahertz Electronics, THz 98, ed. P. Harrison (IEEE, New York 1998), p. 242.
16. P. Haring Bolivar, R. Martini, and H. Kurz, Proc. SPIE Vol. 3828, *Terahertz Spectroscopy and Applications II*, ed. J.M. Chamberlain (SPIE, Bellingham 1999), p. 228.
17. K. Hofbeck, J. Grenzer, E. Schomburg, A.A. Ignatov, K.F. Renk, D.G. Pavel'ev, Yu. Koschurinov, B. Melzer, S. Ivanov, S. Schaposchnikov, and P.S. Kope'ev, Phys. Lett. A **218**, 349 (1996).
18. E. Schomburg, T. Blomeier, K. Hofbeck, J. Grenzer, S. Brandl, I. Lingott, A.A. Ignatov, K.F. Renk, D.G. Pavel'ev, Yu. Koschurinov, B.Ya. Melzer, V.M. Ustinov, S.V. Ivanov, A. Zhukov, and P.S. Kope'ev, Phys. Rev. B **58**, 4035 (1998).
19. R. Scheuerer, E. Schomburg, K.F. Renk, A. Wacker, and E. Schöll, Appl. Phys. Lett. **81**, 1515 (2002).
20. J. Shah, B. Deveaud, T.C. Damen, W.T. Tsang, A.C. Gossard, and P. Lugli, Phys. Rev. Lett. **59**, 2222 (1987).
21. E.R. Brown, T.C.L.G. Sollner, C.D. Parker, W.D. Goodhue, and C.L. Chen, Appl. Phys. Lett. **55**, 1777 (1989).
22. E.R. Brown, J.R. Soderstrom, C.D. Parker, L.J. Mahoney, K.M. Molvar, and T.C. McGill, Appl. Phys. Lett. **59**, 2291 (1989).
23. C. Rauch, G. Strasser, K. Unterrainer, E. Gornik, and B. Brill, Appl. Phys. Lett. **79**, 649 (1997).
24. C. Rauch, G. Strasser, K. Unterrainer, W. Boxleitner, E. Gornik, and A. Wacker, Phys. Rev. Lett. **81**, 3495 (1998).
25. R. Heer, J. Smoliner, G. Strasser, and E. Gornik, Appl. Phys. Lett. **73**, 3138 (1998).
26. R.F. Kazarinov and R.A. Suris, Fiz. Tekh. Poluprovodn. **5**, 797 (1971) [Sov. Phys. Semicond. **5**, 675 (1971)].
27. R.F. Kazarinov and R.A. Suris, Fiz. Tekh. Poluprov. **6**, 148 (1972) [Sov. Phys. Semicond. **6**, 120 (1972)].
28. M. Helm, E. Colas, P. England, F. DeRosa, and S.J. Allen, Jr., Appl. Phys. Lett. **53**, 1714 (1988); M. Helm, P. England, E. Colas, F. DeRosa, and S.J. Allen, Jr., Phys. Rev. Lett. **63**, 74 (1989).
29. S. Luryi, IEEE J. Quantum Electron. QE **27**, 54 (1991).
30. M.P. Shaw, V.V. Mitin, E. Schöll, and H.L. Grubin, *The Physics of Instabilities in Solid State Electron Devices* (Plenum, New York 1992).

31. H. Kroemer, IEEE Spectrum **5**, 47 (1968).
32. L. Esaki and R. Tsu, IBM J. Res. Dev. **14**, 61 (1970).
33. S.A. Ktitorov, G.S. Simin, and V.Y. Sindalovskii, Fiz. Tverd. Tela **13**, 2230 (1971) [Sov. Phys. Solid State **13**, 1872 (1972)].
34. H. Kroemer, *Large-Amplitude Oscillation Dynamics and Domain Suppression in a Superlattice Bloch Oscillator*, cond-mat/0009311.
35. G. Bastard and R. Ferreira, C.R. Acad. Sci. Paris **312 II**, 971 (1991).
36. A.A. Ignatov, K.F. Renk, and E.P. Dodin, Phys. Rev. Lett. **70**, 1996 (1993).
37. F. Löser, M.M. Dignam, Yu.A. Kosevich, K. Köhler, and K. Leo, Phys. Rev. Lett. **85**, 4763 (2000).
38. H. Willenberg, G.H. Döhler, and J. Faist, Phys. Rev. B **67**, 085315 (2003).
39. S.J. Allen, Jr., J.S. Scott, M.C. Wanke, K. Maranowski, A.C. Gossard, M.J. Rodwell, and D.H. Chow, in Proc. NATO Advanced Research Workshop on Terahertz Sources and Systems, Chateau des Bonas, France, 2000, eds. R.E. Miles, P. Harrison, and D. Lippens, NATO Science Series II, Vol. 27 (Kluwer Academic, Dordrecht 2001), p. 3.
40. A. Wacker, S.J. Allen, Jr., J.S. Scott, M.C. Wanke, and A.-P. Jauho, Phys. Status Solidi B **204**, 95 (1997).
41. X.L. Lei, N.J.M. Horing, H.L. Cui, and K.K. Thornber, Appl. Phys. Lett. **65**, 2984 (1994).
42. J. Faist, F. Capasso, D.L. Sivco, C. Sirtori, A.L. Hutchison, and A.Y. Cho, Science **264**, 553 (1994).
43. G. Scamarcio, F. Capasso, C. Sirtori, J. Faist, A.L. Hutchison, D.L. Sivco, and A.Y. Cho, Science **276**, 773 (1997).
44. R. Köhler, R.C. Iotti, A. Tredicucci, and F. Rossi, Appl. Phys. Lett. **79**, 3920 (2001).
45. R. Köhler, A. Tredicucci, F. Beltram, H.E. Beere, E.H. Linfield, A.G. Davies, D.A. Ritchie, R.C. Iotti, and F. Rossi, Nature **417**, 156 (2002).
46. R. Köhler, A. Tredicucci, F. Beltram, H.E. Beere, E.H. Linfield, A.G. Davies, D.A. Ritchie, S.S. Dhillon, and C. Sirtori, Appl. Phys. Lett. **82**, 1518 (2003).
47. M. Rochat, L. Ajili, H. Willenberg, J. Faist, H. Beere, G. Davies, E. Linfield, and D. Ritchie, Appl. Phys. Lett. **81**, 1381 (2002).

Selected List of Symbols

Symbol	Meaning
a	well width
b	barrier width
d	period of superlattice, $d = a + b$
E_0	energy of a reference state in the Wannier–Stark ladder
E_{BO}	field energy from well to well (quantum energy of the Bloch oscillations)
E_{g}	band gap (between the bulk valence and conduction bands)
E_{n}	energies of the Wannier–Stark ladder
e_{i}	peak shift of the Wannier–Stark ladder
F	static electric field
F_{C}	critical electric field
F_ω	prefactor of a harmonically modulated AC field
F^*	field at which the energy step of the Wannier–Stark ladder is identical to the photon energy of the THz field
I_0, I_1	modified Bessel functions
j	current density
k	reciprocal space vector (in a one-dimensional periodic system)

Selected List of Symbols

$\boldsymbol{k}$	reciprocal space vector (in a three-dimensional periodic system)
$\boldsymbol{k}_z$	reciprocal space vector in z direction, along the superlattice transport direction
k_B	Boltzmann constant
L	localization length
m	carrier mass
n	carrier density
n_{well}	sheet carrier density per well
T_1	interband population relaxation time (longitudinal relaxation time)
T_2	polarization relaxation time (transverse relaxation time)
T_2^{inter}	interband polarization relaxation time
T_2^{intra}	intraband polarization relaxation time
T_{12}	delay time in four–wave mixing experiments
T_B	Bloch oscillation period
U	bias voltage
$V(z)$	superlattice potential
v_d	drift velocity
Δ	miniband width
ΔE	gap between first and second miniband
$\epsilon(\omega)$	dynamic dielectric constant
ϵ_∞	high-frequency limit of the dielectric constant
γ	broadening

γ_a	broadening due to acoustic–phonon scattering
γ_{intra}	broadening of the intraband coherence
γ_{LO}	broadening due to optical–phonon scattering
γ_n	broadening due to carrier–carrier scattering
γ_0	background broadening
$\hbar\omega_{\text{LO}}$	longitudinal optical phonon energy
λ	miniband index
μ	mobility
$\varphi_{k_z,\lambda}(z)$	Bloch wave state of an electron in miniband λ
ρ	density matrix
σ	conductivity
τ	a) scattering time b) delay time in pump-probe experiments
$\chi_{\text{loc},\lambda}(z)$	localized state in one of the wells of a superlattice
ω_B	Bloch oscillation angular frequency

Index

above-barrier states, 125, 130
above-gap excitation, 94
absorption spectra, 123
AC conductivity, 164
AC transport, 164
Airy function, 16, 123
anharmonic dispersion, 203
anharmonic oscillations, 74
anticrossing, 122
atom Bloch oscillation, 201
atoms in optical lattices, 201
Auston switch, 88

band transport, 181
band transport models, 151
bare scattering time, 72
basis of eigenstates, 28
Bloch emitter, 212
Bloch oscillations, 4
– in bulk materials, 6, 179
– in other systems, 201
– observation of, 44
Bloch wave, 9, 13, 181
Bloch's theorem, 4
Boltzmann equation, 153
breathing mode, 60, 206
Brillouin zone, 6
broadening caused by Zener tunneling, 137, 138

carrier density, 109
carrier heating, 169
cavity feedback, 215
center-of-mass amplitude, 65
center-of-mass motion, 206
coherence, electronic, 209
coherence, photonic, 210
coherent DC current, 76

coherent electronic effects, 209
coherent Hall effect, 101
coherent quantum transport, 80
conduction band offset, 14
continuous spectrum, 16
Coulomb interaction, 28
coupled double quantum wells, 89
coupling between Bloch oscillations and optical phonons, 113
current resonances, 190
cyclotron energy, 163
cyclotron motion, 100

damping of Bloch oscillations, 103
– due to Zener breakdown, 144
deep-well superlattice, 125
density matrix formalism, 30
dephasing processes, 31
dephasing time, 138
detector application, 211
diagonal density matrix elements, 30
diagonal transitions, 20
differential velocity, 152, 160
disorder, 173
dispersion of superlattice, 12
displacement of wave packet, 57
dissipative transport, 177
distribution function, 153
divergence of the amplitude of Bloch oscillations, 76
domain walls, Bloch oscillations in, 201
domains, of the electric field, 159
doping superlattices, 9
drift transport, 1, 2
Drude transport, 3
dynamically controlled truncation (DCT), 35, 81

electrical spectroscopy, 176
electronic states of a semiconductor superlattice, 11
enlarged well, 156
envelope approximation, 13
envelope function, 13
Esaki–Tsu model, 152
excitation density dependence, 72
exciton binding energy, 22
exciton continuum states, 34
excitonic states, 21

fan chart, 20
Fano coupling, 131, 136
Fano resonances, 23
field-induced delocalization, 123, 127
field-induced localization, 126, 132
four-wave mixing, 36
free-running Bloch emitter, 212
frequency-mixing, 211

gain, 217, 218
gated antenna, 88
gedankenexperiment, 4
graded-gap superlattice, 156
growth techniques, 10
Gunn effect, 2
Gunn oscillator, 2

high-field domain, 159
homogeneous dephasing, 32
hopping transport, 151, 181
hot electron transistor, 176

infrared Drude absorption, 188
inhomogeneous dephasing, 31
instability, 162, 218, 223
interband coherence, 30
interband damping rate, 105
interband dephasing, 105
interband phase relaxation, 39
interface roughness, 155
interface scattering, 177
interminiband gap, 135
intraband coherence, 29
intraband damping rate, 110
intraband damping time, 141
intraband dephasing, 105
intraband dephasing rate, 107

intraband phase relaxation, 39
intraminiband absorption, 190
inverse Bloch oscillator, 77, 190
ionized-impurity scattering, 154

joint density of states, 123
Josephson effect, 77

Kane functions, 17
Kane model, 16
Kronig–Penney model, 11

linear peak shift, 77
Liouville equation, 30
LO phonon frequency, 116
LO phonon scattering, 96
localization, 18
localization length, 19
longitudinal relaxation, 31
low-field mobility, 162, 172

macroscopic dipole moment, 30
magnetic field, absorption spectra in, 134
magnetic field, in transport experiments, 162
magnetic field, THz emission in, 100
metal-organic vapor phase epitaxy (MOVPE), 10
miniband, 9
miniband resonances, 177
miniband width, 48
minigap, 177
mobility, ambipolar, 157
mobility, low-field, 162, 172
mobility, of miniband electrons, 158
molecular-beam epitaxy (MBE), 10
momentum relaxation time, 3

narrow minibands, 171
nearest-neighbor approximation, 18
negative differential conductivity, 160, 180
negative differential velocity, 156, 161, 165, 187
nondiagonal miniband transitions, 124, 130
nonresonant tunneling to continuum states, 124, 130

nonvertical Wannier–Stark ladder transitions, 124, 130

off-diagonal density matrix elements, 30
Ohm's law, 2
one-dimensional Hamiltonian, 16
optical Bloch equations, 31
optical phonons, 113
optical techniques to detect Bloch oscillations, 35
optical waveguide, 204
oscillator strength, 21, 127

phase space filling, 36
phonon resonances, 185
phonon scattering, 154, 177
phonon scattering rate, 171
plasmons, coupling to Bloch oscillations, 68
Poisson equation, 161
polarization interference, 47
polarization wave, 35
pump–probe experiments, 36, 140

quantum beats, 28
quantum cascade laser, 227
quantum computing, 210
quantum mechanical interference, 47

Rabi frequency, 30
relaxation time ansatz, 31
relaxation time approximation, 153
revival of Bloch wave packet, 145
revival of Rydberg wave packet, 145
revival time, 146

scattering time, in bulk SiC, 180
scattering, influence on spatial dynamics, 68
screening, 46
self-induced Shapiro effect, 76
semiconductor Bloch equations, 33
semiconductor superlattices, 9
sequential-tunneling transport, 151, 183
shallow-well superlattices, 128
Shapiro steps, 77, 190
silicon carbide (SiC), 179
spatial amplitude of Bloch wave packet oscillation, 29

spatial dynamics of Bloch oscillations, 29, 53
staggered-alignment superlattice, 10
superconductors, analogy to, 76
superradiance, 213

temperature dependence, of THz emission, 91
terahertz, see THz, 87
thermal saturation, 154, 188
third-order polarization, 36
three-beam four-wave mixing, 38
three-level system, 32
THz field, 88
THz radiation, 87, 212
– of Bloch laser, 216
– of Bloch-oscillating carriers, 89
– other sources, 227
THz spectroscopy, 87
tight-binding model, 18
time-of-flight technique, 166
total amplitude of Bloch oscillations, 58
transient absorption, 36
transistor structures, 157
transmittive electro-optic sampling (TEOS), 41, 55
transverse relaxation, 31
tuning range, 46
two-beam four-wave mixing, 38
two-level system, 30
type I superlattice, 10
type II superlattice, 10

unitary transformation, 28

velocity overshoot, in miniband, 98, 99

Wannier–Stark ladder, 6, 16, 179
– correspondence to negative differential velocity, 186
– fan chart, 20
– modulation, 55
– selection rules for transitions, 21
– transitions, 20
Wannier–Stark resonances, 16
wave guide array, 201
wave packet, 28, 29
wave packet dynamics
– control of, 60

Zener breakdown, 119
Zener coupling, strong limit, 135
Zener resonances, 185

Zener tunneling, 8, 121, 193
– in other systems, 201
– rate, 134

Printing: Saladruck Berlin
Binding: Stürtz AG, Würzburg

Springer Tracts in Modern Physics

147 **Dispersion, Complex Analysis and Optical Spectroscopy**
By K.-E. Peiponen, E.M. Vartiainen, and T. Asakura 1999. 46 figs. VIII, 130 pages

148 **X-Ray Scattering from Soft-Matter Thin Films**
Materials Science and Basic Research
By M. Tolan 1999. 98 figs. IX, 197 pages

149 **High-Resolution X-Ray Scattering from Thin Films and Multilayers**
By V. Holý, U. Pietsch, and T. Baumbach 1999. 148 figs. XI, 256 pages

150 **QCD at HERA**
The Hadronic Final State in Deep Inelastic Scattering
By M. Kuhlen 1999. 99 figs. X, 172 pages

151 **Atomic Simulation of Electrooptic and Magnetooptic Oxide Materials**
By H. Donnerberg 1999. 45 figs. VIII, 205 pages

152 **Thermocapillary Convection in Models of Crystal Growth**
By H. Kuhlmann 1999. 101 figs. XVIII, 224 pages

153 **Neutral Kaons**
By R. Beluševi 1999. 67 figs. XII, 183 pages

154 **Applied RHEED**
Reflection High-Energy Electron Diffraction During Crystal Growth
By W. Braun 1999. 150 figs. IX, 222 pages

155 **High-Temperature-Superconductor Thin Films at Microwave Frequencies**
By M. Hein 1999. 134 figs. XIV, 395 pages

156 **Growth Processes and Surface Phase Equilibria in Molecular Beam Epitaxy**
By N.N. Ledentsov 1999. 17 figs. VIII, 84 pages

157 **Deposition of Diamond-Like Superhard Materials**
By W. Kulisch 1999. 60 figs. X, 191 pages

158 **Nonlinear Optics of Random Media**
Fractal Composites and Metal-Dielectric Films
By V.M. Shalaev 2000. 51 figs. XII, 158 pages

159 **Magnetic Dichroism in Core-Level Photoemission**
By K. Starke 2000. 64 figs. X, 136 pages

160 **Physics with Tau Leptons**
By A. Stahl 2000. 236 figs. VIII, 315 pages

161 **Semiclassical Theory of Mesoscopic Quantum Systems**
By K. Richter 2000. 50 figs. IX, 221 pages

162 **Electroweak Precision Tests at LEP**
By W. Hollik and G. Duckeck 2000. 60 figs. VIII, 161 pages

163 **Symmetries in Intermediate and High Energy Physics**
Ed. by A. Faessler, T.S. Kosmas, and G.K. Leontaris 2000. 96 figs. XVI, 316 pages

164 **Pattern Formation in Granular Materials**
By G.H. Ristow 2000. 83 figs. XIII, 161 pages

165 **Path Integral Quantization and Stochastic Quantization**
By M. Masujima 2000. 0 figs. XII, 282 pages

166 **Probing the Quantum Vacuum**
Pertubative Effective Action Approach in Quantum Electrodynamics and its Application
By W. Dittrich and H. Gies 2000. 16 figs. XI, 241 pages

167 **Photoelectric Properties and Applications of Low-Mobility Semiconductors**
By R. Könenkamp 2000. 57 figs. VIII, 100 pages

168 **Deep Inelastic Positron-Proton Scattering in the High-Momentum-Transfer Regime of HERA**
By U.F. Katz 2000. 96 figs. VIII, 237 pages

Springer Tracts in Modern Physics

169 **Semiconductor Cavity Quantum Electrodynamics**
By Y. Yamamoto, T. Tassone, H. Cao 2000. 67 figs. VIII, 154 pages

170 **d–d Excitations in Transition-Metal Oxides**
A Spin-Polarized Electron Energy-Loss Spectroscopy (SPEELS) Study
By B. Fromme 2001. 53 figs. XII, 143 pages

171 **High-T_c Superconductors for Magnet and Energy Technology**
By B. R. Lehndorff 2001. 139 figs. XII, 209 pages

172 **Dissipative Quantum Chaos and Decoherence**
By D. Braun 2001. 22 figs. XI, 132 pages

173 **Quantum Information**
An Introduction to Basic Theoretical Concepts and Experiments
By G. Alber, T. Beth, M. Horodecki, P. Horodecki, R. Horodecki, M. Rötteler, H. Weinfurter, R. Werner, and A. Zeilinger 2001. 60 figs. XI, 216 pages

174 **Superconductor/Semiconductor Junctions**
By Thomas Schäpers 2001. 91 figs. IX, 145 pages

175 **Ion-Induced Electron Emission from Crystalline Solids**
By Hiroshi Kudo 2002. 85 figs. IX, 161 pages

176 **Infrared Spectroscopy of Molecular Clusters**
An Introduction to Intermolecular Forces
By Martina Havenith 2002. 33 figs. VIII, 120 pages

177 **Applied Asymptotic Expansions in Momenta and Masses**
By Vladimir A. Smirnov 2002. 52 figs. IX, 263 pages

178 **Capillary Surfaces**
Shape – Stability – Dynamics, in Particular Under Weightlessnes
By Dieter Langbein 2002. 182 figs. XVIII, 364 pages

179 **Anomalous X-ray Scattering for Materials Characterization**
Atomic-Scale Structure Determination
By Yoshio Waseda 2002. 132 figs. XIV, 214 pages

180 **Coverings of Discrete Quasiperiodic Sets**
Theory and Applications to Quasicrystals
Edited by P. Kramer and Z. Papadopolos 2002. 128 figs., XIV, 274 pages

181 **Emulsion Science**
Basic Principles. An Overview
By J. Bibette, F. Leal-Calderon, V. Schmitt, and P. Poulin 2002. 50 figs., IX, 140 pages

182 **Transmission Electron Microscopy of Semiconductor Nanostructures**
An Analysis of Composition and Strain State
By A. Rosenauer 2003. 136 figs., XII, 238 pages

183 **Transverse Patterns in Nonlinear Optical Resonators**
By K. Staliūnas, V. J. Sánchez-Morcillo 2003. 132 figs., XII, 226 pages

184 **Statistical Physics and Economics**
Concepts, Tools and Applications
By M. Schulz 2003. 54 figs., XII, 244 pages

185 **Electronic Defect States in Alkali Halides**
Effects of Interaction with Molecular Ions
By V. Dierolf 2003. 80 figs., XII, 196 pages

186 **Electron-Beam Interactions with Solids**
Application of the Monte Carlo Method to Electron Scattering Problems
By M. Dapor 2003. 27 figs., X, 110 pages

187 **High-Field Transport in Semiconductor Superlattices**
By K. Leo 2003. 164 figs., XIV, 242 pages